# Microwave Plasma Sources and Methods in Processing Technology

# Microwave Plasma Sources and Methods in Processing Technology

*Ladislav Bárdoš and Hana Baránková*
*Uppsala University, Uppsala, Sweden*

WILEY

IEEE PRESS

This edition first published 2022


Published by John Wiley & Sons, Inc., Hoboken, New Jersey.
Published simultaneously in Canada.



*Library of Congress Cataloging-in-Publication Data*
Names: Bardos, Ladislav, 1947- author. | Barankova, Hana, 1951- author.
Title: Microwave plasma sources and methods in processing technology / Ladislav Bardos, Hana Barankova, Uppsala University, Uppsala, Sweden.
Description: Hoboken, New Jersey : John Wiley & Sons, 2022. | Includes bibliographical references and index.
Identifiers: LCCN 2021031913 (print) | LCCN 2021031914 (ebook) | ISBN 9781119826873 (hardback) | ISBN 9781119826880 (pdf) | ISBN 9781119826897 (epub) | ISBN 9781119826903 (ebook)
Subjects: LCSH: Electromagnetism.
Classification: LCC QC760 .B267 2022 (print) | LCC QC760 (ebook) | DDC 660/.044--dc23
LC record available at https://lccn.loc.gov/2021031913
LC ebook record available at https://lccn.loc.gov/2021031914

Cover image: © Image courtesy of the authors
Cover design by Wiley

Set in 9.5/12.5pt STIXTwo Text by Integra Software Services, Pondicherry, India

10 9 8 7 6 5 4 3 2 1

M 03M000030000004758 040222

# Contents

# Foreword from the Authors

After our long research and development activities in Plasma science we can state that ***the plasma is a well controlled environmentally-friendly medium utilizing electric power and enabling the very-high temperature processes and low temperature reactions not available in other methods.*** This book is, therefore, intended to help the readers interested in the non-thermal gas discharge plasmas and their applications. The content is focused on the microwave-generated plasmas, which have specific properties due to relatively high frequency and which can be used in a number of non-conventional applications. Based on our experience, from university teaching and from communications with the industry people, no broad understanding exists of the microwave power and the microwave plasma systems. People know microwave ovens, but mostly for the heating of meals, and without understanding the principles and specifications of the microwave power. Little knowledge exists about different plasma systems and applications related to microwave power, even though there is apparently growing interest in this technology, mainly for new plasma chemical processes. A motivation for writing this book is based on the authors' long experience with design and applications of several non-conventional systems and applications, which might stimulate readers when furthering their knowledge and when developing new systems.

The content of this book is composed of five basic parts, i.e. chapters. Chapter 1 is devoted to the microwave techniques and power systems from introducing their short history to the descriptions of individual parts and components in the microwave power lines, which are used in laboratory experiments and many industrial devices. Besides the microwave communication and the radar techniques, a lot of original systems have been developed for heating the plasma in fusion test reactors (Tokamaks, Stellarators, and Magnetic Mirrors), as well as in particle accelerators. Moreover, in the last two decades, interest has rapidly grown in microwave plasma systems working at atmospheric and higher pressures, in gases and inside liquids. Interesting new applications and trends are described in Chapter 6.

We have no intention to write a textbook and start with explanations of basic microwave theories based on Maxwell equations. Excellent comprehensive monographies have been written about microwave engineering with all theories, simulation models, and details, including use of microwaves for generation of the plasma discharges, see e.g. Refs. [1–7]. However, to make it easier for the book readers without experience in the field of microwave engineering, we introduce the microwave systems in a simple way. We illustrate and describe the most important microwave components and show at least the most important expressions, which can help in greater understanding of the principles, functions, and applications of the microwave components. For these purposes we have also created a large number of original illustrations. Therefore, the text is frequently accompanied by schematic pictures, diagrams, and photographs. Such an approach was promoted and recommended long ago by the "Teacher of the nations," Johannes Amos Comenius (Komensky in Czech) in his widely translated book "Orbis Sensualium Pictus," issued in 1658. A copy of this book is available for example in the Hungarian library at the web address https://library.hungaricana.hu/hu/view/RMK_I_1091-RM_I_8r_0547/?pg=0&layout=s. Moreover, at the end of Chapter 1 we have included Part 1.5 describing the microwave oven. This is because each oven represents a smartly engineered microwave system, which can help in the understanding of the basic principles and components in the microwave technology introduced and described in Chapter 1.

After an explanation of the functions of microwave components in the typical power lines for the microwave plasma generation, in Chapter 2, we describe the fundamentals of the gas discharge plasma and differences between plasmas generated by different kinds of the power, with particular emphasis to the microwave power. Chapter 3 is devoted to explanations of interactions of plasmas with solid surfaces and gases, mainly at reduced and low pressures. Used explanations are simplified and limited to basic expressions and equations necessary for the understanding of the processes in the plasma and those caused by the plasma, as described in later chapters. Parts of the texts and some illustrations in Chapters 2 and 3 are used in the authors' courses for university students, short tutorials at companies and, since 1997, in annual courses for the Society of Vacuum Coaters (www.svc.org) in the United States.

Chapter 4 focuses on different microwave plasma systems, including novel and non-conventional ones developed and laboratory tested in different processing applications at reduced pressures. Chapter 5 is devoted to the microwave plasma systems at atmospheric and higher pressures, including plasmas inside liquids and plasma interactions with the combustion flames. Chapter 6 describes some new applications and trends in microwave plasmas, with short opinions and expectations on future perspectives of the microwave plasma and its applications. Chapter 7 contains appendices with description of symbols, abbreviations, units and values used in the individual chapters.

As mentioned, this book has no ambition to become a "handbook" or a "textbook". We have written an "easy" text to inspire the readers and raise their interest in further studies and designs of novel systems, as well as to help readers in their experimental works with the microwave plasma and the microwave plasma-assisted applications.

## References

[1] D. M. Pozar: "*Microwave engineering*", 2nd Ed., John Wiley & Sons Inc., New York, 1996.
[2] J. C. Slater: "*Microwave electronics*", D. Van Nostrand Company Inc., New York, 1951.
[3] T. Moreno: "*Microwave transmission design data*", Artech House Microwave Library, Boston, MA, 1989. ISBN 089006346X, 9780890063460.
[4] L.S. Polak and Yu.A. Lebedev, eds.: "*Plasma Chemistry*", Cambridge International Science Publishing, Cambridge, UK, 1998. ISBN 1898326223, 9781898326229.
[5] M. Moisan and J. Pelletier, eds.: "*Microwave Excited Plasmas Vol. 4*", 1st Ed., Elsevier Science, Amsterdam, Netherlands, 1992.
[6] Yu.A. Lebedev, ed.: "*Microwave discharges: fundamentals and applications*", Yanus-K, Moscow, 2001. ISBN 5-8037-0066–5.
[7] O.A. Popov, ed.: "*High density plasma sources*", Noyes Publications, Park Ridge, NJ, 1995. ISBN 0-8155-1377–1.

# 1

# Basic Principles and Components in the Microwave Techniques and Power Systems

## 1.1 History in Brief – From Alternating Current to Electromagnetic Waves and to Microwaves

The greatest discoveries and developments connected with the great names in the field of alternating current (AC) and related systems were already dated in the early nineteenth century, see Ref. [1.1]. However, an important basic invention was the battery, a source of electricity, disclosed by the Italian scientist Alessandro Volta in 1799. This simple source of a direct current (DC) allowed many important experiments with electricity. In 1820, Danish physicist Hans Christian Ørsted discovered an effect of electricity on the magnetic field and his findings were confirmed by the experiments of French physicists Andre-Marie Ampere and François Arago. The parallel wires with DC current visibly attracted or repelled each other according to the mutual current directions. However, in 1830, English scientist Michael Faraday discovered the ability to generate electricity by moving magnets and the corresponding principle of the electric induction. These new effects based on the electric induction inspired the Serbian-American electrical and mechanical engineer Nikola Tesla and led to his inventions of an alternating current generator which used a rotating magnetic field, the Tesla coil, the transformation of AC voltages to high voltages or vice-versa, as well as other inventions patented at the end of 1887. Besides his fundamental inventions, Tesla is considered a pioneer in radar technology, X-ray technology, and remote control.

In 1860, Scottish scientist James Clerk Maxwell proposed the electromagnetic disturbances longer than infrared (IR) radiation. He explained theoretically how electric and magnetic fields can form electromagnetic waves and developed a theory known as Maxwell's equations, see Ref. [1.2]. In 1888, the German physicist Heinrich Hertz used an electric spark arrangement and a simple capacitor made from a Leyden jar for the generation of electromagnetic waves, see Ref. [1.3]. He was the first person who was able to transmit and receive radio waves. That is why the frequency unit, 1 Hz = 1/s, is called Hertz.

*Microwave Plasma Sources and Methods in Processing Technology*, First Edition.
Ladislav Bárdoš and Hana Baránková.

In 1897, an application of Hertzian waves was patented for long-distance radio telegraph communication by Italian electrical engineer Guglielmo Marconi. Radio waves are only part of electromagnetic waves used in radio communications. The whole spectrum of electromagnetic waves is graphically illustrated in Figure 1.1. The spectrum can be roughly divided into the following wavelengths ($\lambda$):

- Gamma ($\gamma$) rays $\lambda \approx$ 1–100 pm (picometer)
- X-rays $\lambda \approx$ 100 pm–100 nm
- Ultraviolet (UV) light $\lambda \approx$ 100–300 nm
- Visible light $\lambda \approx$ 350–750 nm
- IR $\lambda \approx$ 750 nm–1 mm
- Microwaves ($\mu$) $\lambda \approx$ 1 mm–1 m
- Radio waves $\lambda \approx$ 1 m–100 km

The electromagnetic waves are moving with the speed of light ($c = 3 \times 10^8$ m/s) and the corresponding frequencies can be calculated from the expression $f = c/\lambda$. This gives the following approximate frequency values:

- Gamma ($\gamma$) rays $f \approx 3 \times (10^{20}–10^{18})$ Hz
- X-rays $f \approx 3 \times (10^{18}–10^{15})$ Hz
- UV light $f \approx 3 \times 10^{15}–10^{15}$ Hz
- Visible light $f \approx 9 \times 10^{14}–4 \times 10^{14}$ Hz
- IR $f \approx 4 \times 10^{14}–3 \times 10^{11}$ Hz
- Microwaves ($\mu$) $f \approx 3 \times 10^{1}–3 \times 10^{8}$ Hz
- Radio waves $f \approx 3 \times 10^{8}–3 \times 10^{4}$ Hz

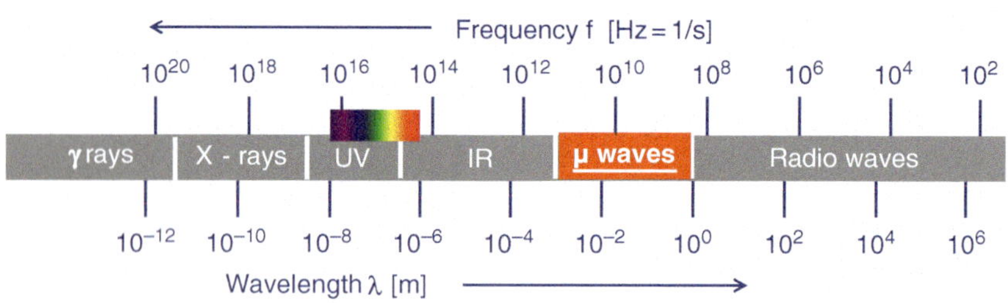


**Figure 1.1** Graphical illustration of the frequencies and wavelengths in the spectrum of the electromagnetic waves. An optical spectrum is inserted between ultraviolet and infrared waves. The part of the visible light is shown in spectral colors.

In case you need to recall what units are used in the high and very high frequencies, see the following list:

- kHz $10^3$ Hz (Kilohertz)
- MHz $10^6$ Hz (Megahertz)
- GHz $10^9$ Hz (Gigahertz)
- THz $10^{12}$ Hz (Terahertz)
- PHz $10^{15}$ Hz (Petahertz)
- EHz $10^{18}$ Hz (Exahertz)
- ZHz $10^{21}$ Hz (Zettahertz)
- YHz $10^{24}$ Hz (Yottahertz).

The ideas presented by Tesla along with many ongoing experiments with electromagnetic waves in many countries led to the development of electromagnetic detection of objects, i.e. radar, see the condensed history in Ref. [1.4]. In 1936, American Robert M. Page and his group developed a device for detecting planes at distances up to 40 km. Four years later, the U.S. Navy named such systems “Radar”, meaning Radio Detection and Ranging. Almost simultaneously, Robert Watson-Watt in England developed a radar device according to his patent filed in 1935. In Germany after 1933, Rudolf Kühnhold and Hans Erich Hollmann developed radar systems, one of which was called Würzburg. Further technical developments of radars were accelerated considerably during the World War Two (WWII).

The detection of moving objects by reflected waves can be performed only by electromagnetic waves proceeding in straight lines and directions selected by the operator. Because this was fulfilled best by using microwaves, all developments around radar were closely connected with the developments of suitable microwave generators. In the document, see Ref. [1.5], issued by AT&T company in 1947 about building a 7.5-cm (3 inches) wavelength communication system working at f = 4 GHz between New York and Boston they wrote: “Our microwaves are more closely related to the light waves than they are to the waves used for broadcasting, and this accounts for the fact that they travel in straight lines from one point to the next, can be reflected by smooth bright surfaces concentrated into a point source by a lens.”

## 1.2 Microwave Generators

The microwave wavelengths can be roughly considered to be between 1 mm and 1 m, which corresponds to the frequency range between 300 GHz and 300 MHz. There are about 13 different frequency intervals (bands) designed for different applications, particularly for communications and military systems. The focus in this book will be in the L-band (1–2 GHz) and mainly in the S-band (2–4 GHz), which are used in the powering of systems like microwave ovens, microwave medical instruments, and the generation of the microwave plasmas. The microwaves in this frequency region are also noted as "centimeter" waves. The frequency 2.4±0.5 GHz in such microwave systems is called an "industrial frequency" because, even at a power of hundreds of kW, these systems must not disturb the communication channels.

Several types of microwave power generators exist: travelling wave tube (TWT), klystron, magnetron, or gyrotron, see Ref. [1.6]. All these generators utilize electric and magnetic fields to drive and control motions of electrons and to produce microwave frequencies with the desired power level. Microwave power generators are designed for different frequencies and output powers. Different generators also exhibit different efficiencies (the ratio of the outlet microwave power to the applied electric input) from about 15% to as high as 95%. There are differences in the construction, power control and the maximum output power between "classical" vacuum tube-based generators and those using solid-state components, like microwave diodes. A simple reason is the limited voltage applicable in the solid-state semiconductor components and corresponding power limits. This book is dealing with the magnetron tubes, which differs from the magnetron sputtering diodes for film coatings.

Most of the microwave sources for generation of gas discharge plasmas are equipped with a magnetron generator, which is an evacuated tube used, e.g. in microwave ovens (see Section 1.5). Such a tube is a vacuum diode with oscillating electrons generating the microwave power, typically from 0.5–15 kW. Magnetron tube generators built for 2.4 GHz (13 cm waves) are relatively cheap, and their power efficiencies are about 60%. The construction of typical magnetron is schematically shown in Figure 1.2.

The anode is a hollow cylinder with an even number of anode segments. The open trapezoidal shaped areas between each of the segments represent resonant cavities serving as tuned circuits. Their shapes and dimensions determine the output frequency of the tube.

The cathode filament (heater) enhances the electron emission from the cathode located in the center of the magnetron tube coaxially with the anode.

The magnetic field provided by strong permanent magnets around the magnetron and the magnetic field is parallel with the axis of the cathode. The magnetic field turns the electrons to circulate along the anode segments.

The antenna is a probe or loop connected to the anode and extends into one of the tuned cavities. The antenna transmits the power generated by the resonances.

The magnetron utilizes oscillation motions of electrons under combined effect of the electric and magnetic field. After switching on the high voltage

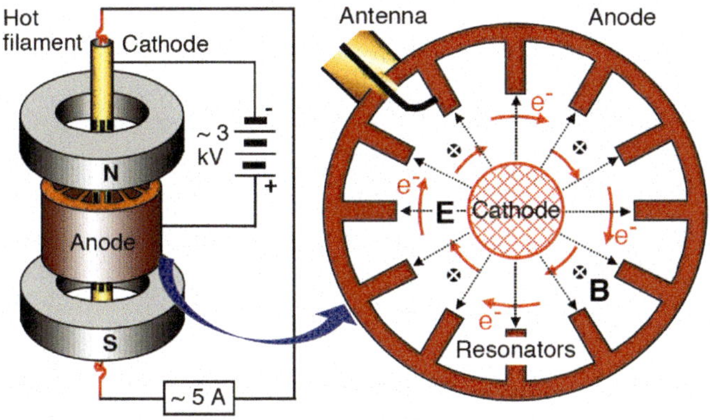


**Figure 1.2** Construction of the magnetron tube consists of the anode, the hot filament/cathode, the antenna, and magnets. The anode and cathode arrangement are in an evacuated tube arranged with an axial magnetic field. The **E** and **B** are vectors of the electric and magnetic fields. The arrows in the segmented anode show circulating electrons.

(several kV) between the anode (grounded) and the cathode, the electrons emitted from the cathode are accelerated to the anode and turned by the magnetic field to circular paths. Such motions of the electrons cause resonances in the anode segments dimensioned to resonate at the microwave frequency. There must be a minimum electric field necessary to energize electrons for reaching the anode and starting the resonances. The power generated by the magnetron is guided to a coaxial output of the magnetron with an antenna being the central conductor. More details about the electric circuit powering the magnetrons are introduced in Section 1.5.

## 1.3 Waveguides and Electromagnetic Modes in Wave Propagation

There are many transmission systems of electromagnetic waves. The low-frequency power with a wavelength longer than 1 m (frequency below 300 MHz) can be transported by two parallel wires or by coaxial electric conductors (coaxial cables). Such lines, called transmission lines, transport electric power in forms of currents and voltages. Their transmission efficiency depends on the line resistance, capacitance, inductance, and shunt resistance between conductors. However, transmission of centimeter or shorter waves by such lines, particularly at higher powers, can cause high power losses due to radiation and attenuation of the wave transport. It is better to “close” the short waves into hollow structures, e.g. as conductive tubes made from brass or copper (Cu) alloys with aluminum (Al) and having inner dimensions comparable with the wavelength. In these waveguides, the power is transported by the moving waves of electric and magnetic fields.

In ordinary electric transmission lines, e.g. in the coaxial lines or parallel wires, transversal electro-magnetic (TEM) is the typical mode for transporting the electric power. Here, electric and magnetic fields only have components perpendicular to the axis of the transfer, and they have zero amplitudes along the line (z-axis in the Cartesian coordinates). The electric field is between the conductors, and the magnetic field is around the conductors transporting the current. On the contrary, in the waveguides, the waves are transported in certain modes, either transversal electric (TE) or transversal magnetic (TM), but TEM modes cannot propagate here. The TE modes have no electric component along the wave transport (vector $\boldsymbol{E}_z = 0$) and only the magnetic component is non-zero along the transport (vector $\boldsymbol{H}_z \neq 0$). In other words, in the TE modes, the waves are propagating only with an electric field perpendicular to the propagation. The TM modes are the opposite for the $\boldsymbol{H}$ and $\boldsymbol{E}$ vectors in that $\boldsymbol{H}_z = 0$, $\boldsymbol{E}_z \neq 0$. The waveguide power lines in the generation of microwave plasmas use mostly air-filled rectangular or circular waveguides with the *dominant modes*. The dominant modes are those where the waves propagate at the lowest frequency (longest wavelength), i.e. close to the *cut-off frequency* ($f_{coff}$) for the used waveguide (see Figure 1.3). Dominant modes exhibit the lowest

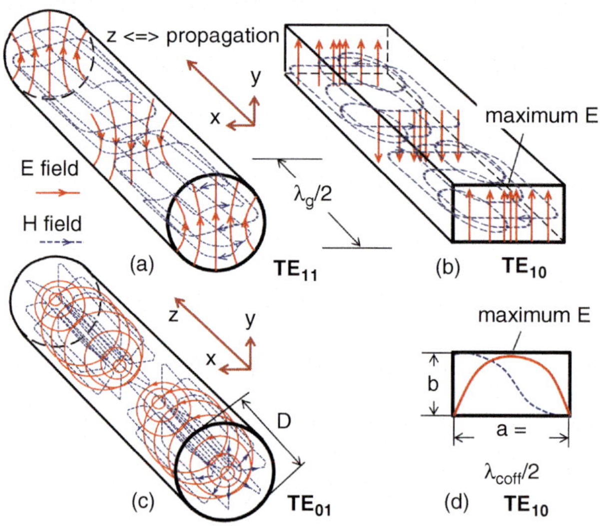


**Figure 1.3** The electric and magnetic fields in the waveguides. (a) Dominant mode ($TE_{11}$) in the circular waveguide. (b) Dominant mode ($TE_{10}$) in the rectangular waveguide. (c) Mode $TE_{01}$ in the circular waveguide. (d) Intensity of the electric field in the $TE_{10}$ mode. The wavelength $\lambda_g$ is the length of propagating wave, the $\lambda_{coff}$ is the cut-off wavelength.

attenuation of the waves and there is also a low chance for any other mode to appear and interfere with the wave propagation. Simplified illustrations of the dominant waveguide modes in circular ($TE_{11}$) and rectangular ($TE_{10}$) waveguides are shown in Figure 1.3.

The indexes m and n in the modes $TE_{mn}$ denote numbers of half wavelengths ($\lambda/2$) of the electric field component (m) and the magnetic field component (n) in the cross-section of the waveguide. The difference between the modes is clear when comparing the dominant mode $TE_{11}$ (Figure 1.3a) with the mode $TE_{01}$ (Figure 1.3c) in the circular waveguide with the inner diameter D. For the $TE_{10}$ mode in the rectangular waveguide (inner dimensions a and b), it means one transversal component of the electric field ($m = 1$). The corresponding profile of the intensity of the E field is a simple "hill" (one-half of the cut-off wavelength), see Figure 1.3d. The detailed knowledge about distributions of electric and magnetic fields in the waveguides is important for the right designs of the reactors intended for the microwave generation of the plasmas.

### 1.3.1 The Cut-off Frequency and the Wavelength in Waveguides

Waveguides have "disperse" properties, which means they depend on the frequency of waves to be transported. Dimensions of the waveguides limit the frequencies (and wavelengths) of the waves for the transport. If the wave frequency f is higher than the cut-off frequency for different modes of the waves, then all these modes can propagate in the waveguide. If the frequency f is less than all cut-off frequencies, then none of these modes can propagate inside. As noted, the lowest wave attenuation can be obtained for the longest ("upper") usable wavelength (the lowest frequency), the length of which depends on the physical dimensions of the waveguide and on the propagation mode. Therefore, the best mode is the dominant mode with the frequency near the characteristic "lower" cut-off frequency of the given waveguide. The dominant modes are also used in most of the applications of the plasma systems. This cut-off frequency ($f_{coff}$) for the $TE_{mn}$ and $TM_{mn}$ modes in the air-filled rectangular waveguides can be expressed as the following:

$$2\, f_{coff} = c\left[(m/a)^2 + (n/b)^2\right]^{1/2} \tag{1.1}$$

and the cut-off frequency for the dominant mode $TE_{10}$ ($m = 1$, $n = 0$) will be the following:

$$f_{coff} = c/2a = 1.5/a10^{10}\,\text{Hz} \tag{1.2}$$

where c = 3 x $10^{10}$ cm/s is the velocity of light in vacuum (air) with the cross-dimension a (in the x-axis) in centimeters. The corresponding cut-off wavelength is expressed as the following:

$$\lambda_{coff} = 2a \tag{1.3}$$

For the maximum power transfer in the rectangular waveguides, it is always a = 2b. For example, with a = 8.6 cm, b = 4.3 cm for the microwave frequency of 2.35 GHz), the cut-off frequency is $f_{coff}$ = 1.74 GHz. The dimensions a and b given above are almost identical with the Electronic Industry Association (EIA) standard waveguides WR340. A corresponding expression for the cut-off frequency of the dominant mode $TE_{11}$ in the circular waveguide with diameter D is the following:

$$f_{coff} = 0.293\, c/D \approx 0.88/D\ 10^{10}\text{Hz} \tag{1.4}$$

For the cut-off frequency of the rectangular waveguide, the diameter D in circular waveguide should correspond to D ≈ 1.5 a /0.88 = 1.7 a. Therefore, the cross-dimension of the circular waveguides is usually considerably larger than in the rectangular waveguides. Despite their simpler fabrication, circular waveguides consume more metal and are heavier. For the cut-off wavelength in the circular waveguide, it is the following:

$$\lambda_{coff} = \pi D/\ 1.84 = 1.7\ D \tag{1.5}$$

The direction vector of the wave propagation in the waveguides is not parallel with the z-axis. It has an acute angle to the z-axis, and the waveguide transport is based on the wave reflections between the opposite conductive walls and is shown schematically in Figure 1.4.

The wavelength in the waveguide differs from the free space, and it can be expressed as the following:

$$\lambda_g = \lambda / \left[1 - (\lambda / \lambda_{coff})^2\right]^{1/2} \tag{1.6}$$

The wavelength in the waveguide is longer than that in the free space ($\lambda_g > \lambda$) because $\lambda < \lambda_{coff} = 2a$ (for $TE_{10}$), and the denominator in Eq. (1.6) is then smaller than one. The velocity of the wave propagation is slower than in free space ($v_g < v_c = c$). For 2.35 GHz, the free-space wavelength is λ = 30/2.35 = 12.77 cm; for 2.4 GHz, it is 12.5 cm. In the rectangular waveguide with a = 8.6 cm, the wavelength $\lambda_g \approx 19$ cm.

Figure 1.4 shows that the velocity of the wave power in z-direction is slower than the propagation of the wave (here it is velocity of light c). The power

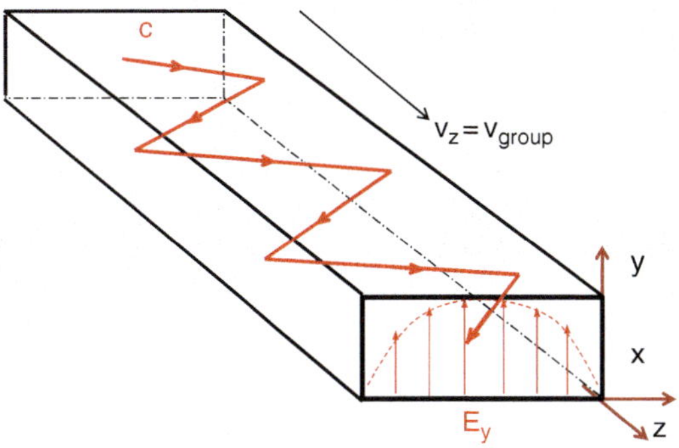


**Figure 1.4** Schematic description of the reflection of the propagating $TE_{10}$ wave in a lossless waveguide. Value ($v_{group}$) is the velocity of propagating the wave power.

carried by the wave in the waveguide is transported at a velocity called group velocity, which can be expressed as the following:

$$v_z = v_{group} = c\left[1-\left(f_{coff}/f\right)^2\right]^{1/2} \tag{1.7}$$

Therefore, $v_{group}$ is less than c. Also, the nodes of the wave in the waveguide, where the wave changes its phase, can propagate virtually at phase velocity ($v_{phase} > c$), higher than the velocity of light. Of course, the phase velocity doesn't transfer the energy. A simple relation exists between the group and the phase velocities:

$$v_{group} \cdot v_{phase} = c^2 \tag{1.8}$$

### 1.3.2 Waveguides Filled by Dielectrics

The velocity of wave propagation introduced above in "empty" or "lossless" waveguides (empty or filled with air) is equal to the speed of light $c = 3 \times 10^8$ m/s (more exactly, 2.998 $10^8$ m/s). Light is also a kind of an electromagnetic wave (see Figure 1.1). There is an important relation $c = (\mu_0\, \varepsilon_0)^{-1/2}$, where the vacuum permeability $\mu_0 = 1.256 \times 10^{-6}$ H/m $= 4\pi\ 10^{-7}$ H/m, and the vacuum permittivity is $\varepsilon_0 = 8.85 \times 10^{-12}$ F/m. (Henry is represented as $H = kg\ m^2\ s^{-2}\ A^{-2}$, and Farad is represented as $F = A^2 s^4\ kg^{-1}\ m^{-3}$). When the waveguide is filled by certain dielectrics, the velocity of the wavelength propagation $c_d$ (as well as the velocity of light) is slower than in the vacuum:

$$c_d = (\mu\varepsilon)^{-1/2}, \text{ where } \mu = \mu_0\mu_r \text{ and } \varepsilon = \varepsilon_0\,\varepsilon_r \tag{1.9}$$

Here, $\mu_r$ and $\varepsilon_r$ are relative permeability and relative permittivity, respectively, of the used dielectric. In a vacuum and in air, they are $\mu_r = \varepsilon_r = 1$ and $c_d$

= c. For the rectangular waveguide filled with a dielectric, the equations (1.1) and (1.3) for the cut-off frequency ($f_{coff}$) and the cut-off wavelength ($\lambda_{coff}$) for $TE_{10}$ should be rewritten as the following:

$$2\,f_{coff} = (\mu\varepsilon)^{-1/2}\left[(m/a)^2 + (n/b)^2\right]^{1/2} = c(\mu_r\varepsilon_r)^{-1/2}\left[(m/a)^2 + (n/b)^2\right]^{1/2} \quad (1.10)$$

$$\lambda_{coff} = 2(\mu_r\varepsilon_r)^{1/2}a \quad (1.11)$$

Because all dielectrics have a relative permittivity of $\varepsilon_r \geq 1$ and a relative permeability of $\mu_r \approx 1$, the cut-off wavelengths for dielectric-filled waveguides will be longer than in the empty waveguides. Considering Eq. (1.2) for the dominant mode $TE_{10}$ (m = 1, n = 0) in an empty waveguide, we have $f_{coff} = c/2a = 1.5/a\ 10^{10}$ Hz; for the same waveguide filled by PTFE, known as Teflon or $[(C_2F_4)_n]$ (with $\varepsilon_r = 2$ and $\mu_r = 1$), it will be $f_{coff} = c_d/2a = 1.5/1.4a\ 10^{10}$ Hz. The cut-off frequency will be 1/1.4 = 0.7 times smaller. Therefore, the dielectric filled waveguide allows lower wave frequency (longer waves) to propagate. However, the dielectric losses will cause power losses and attenuation of the waves. Therefore, it is a bad idea to use dielectric filled waveguides to avoid the involvement of gases or for vacuum sealing of the waveguide parts. For such purpose, vacuum windows in the waveguides can be designed in a different way, see Section 1.4.5.

### 1.3.3 Wave Impedance and Standing Waves in Waveguides

For the wave propagation in any uniform medium, finding the characteristic impedance is possible. In a medium characterized by the permeability $\mu = \mu_0\ \mu_r$ and permittivity $\varepsilon = \varepsilon_0\ \varepsilon_r$, the impedance Z is given as follows:

$$Z = Z_0(\mu/\varepsilon)^{1/2} = (\mu_0/\varepsilon_0)^{1/2}(\mu_r/\varepsilon_r)^{1/2} = 120\pi(\mu_r/\varepsilon_r)^{1/2} = 376.73(\mu_r/\varepsilon_r)^{1/2}\ \text{Ohms} \quad (1.12)$$

where $Z_0 = (\mu_0/\varepsilon_0)^{1/2}$ or 120 $\pi$ or 376.73 Ohms is the impedance of free space (vacuum, air).

When the wave propagates in an "endless" waveguide in TE or TM modes, expressing the wave impedances $Z_{TE}$ and $Z_{TM}$ by the cut-off frequencies is possible:

$$Z_{TE} = Z_0\left[1 - (f_{coff}/f)^2\right]^{1/2} \quad (1.13)$$

$$Z_{TM} = Z_0 \left[1 - \left(f_{coff}/f\right)^2\right]^{1/2} \quad (1.14)$$

For example, in the lossless rectangular waveguide with the dominant mode $TE_{10}$, at the microwave frequency of 2.35 GHz and the cut-off frequency of $f_{coff}$ = 1.74 GHz, calculated from Eq. (1.2), the wave impedance will be $Z_{TE10} = 377/[1-(1.74/2.35)^2]^{1/2} \approx 561$ Ohms.

Regardless of whether the end of the waveguide is opened or closed by certain barriers from any material, part of the propagating wave can return, superimpose with the forwarding wave and form so-called standing wave, see Figure 1.5.

The amplitude of a reflected wave naturally depends on the material terminating the waveguide because certain parts of the wave can be absorbed, and the reflected wave can acquire different phases with respect to the incident wave. Some materials can absorb the wave power almost entirely and transfer it to the heat. This is the case in microwave ovens, where the microwave power at a frequency around 2.4 GHz is efficiently absorbed in water. Because meals usually contain a lot of water, the microwave power can be used for their heating. Some materials have a higher absorption, e.g. ferrites, carbon, or silicon. Such materials can be used as the waveguide loads (examples will be described in Section 1.4.7). The microwave heating is now applied in a number of applications, see, e.g. Refs. [1.7, 1.8]. However, the absorption ability can have undesired consequences, e.g. by melting of glass due to power interacting with the silicon in the silicon dioxide ($SiO_2$) material.

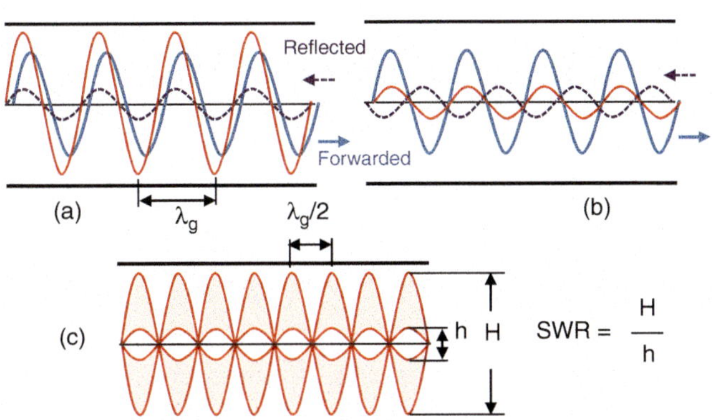


**Figure 1.5** Formation of the standing waves by waves moving in the waveguide. (a) The reflected wave is in a phase similar to the forwarded wave. (b) The reflected wave is in a position close to the opposite phase to the forwarded wave. (c) The resulting "standing" waves.

The resulting standing wave ratio (SWR), i.e. the ratio between highest and lowest part of the standing wave amplitude (measured usually by the voltage standing wave ratio or VSWR), resulted from the interaction between forwarded, and reflected waves can be used for measurements of the impedance terminating the waveguide. According to Figure 1.5, the SWR ratio is SWR = H/h. It is then useful to define the percentage of the wave reflection ($W_r$) as the following:

$$W_r[\%] = 100\,(\mathrm{SWR} - 1)^2 / (\mathrm{SWR} + 1)^2 \tag{1.15}$$

When SWR = 1, then $W_r$ = 0%, and the wave is not reflected. When SWR = 0, then $W_r$ = 100%, and the wave is fully reflected. This is the case when the waveguide is terminated by an electrically conductive short (often shaped as a movable "shorting plunger"). For practical uses, the defined so-called reflection coefficient (**Γ**) is a complex value, and its scalar form |Γ| can be expressed using the SWR in the following way:

$$|\Gamma| = (\mathrm{SWR} - 1)/(\mathrm{SWR} + 1) \text{ or } \mathrm{SWR} = (1 + |\Gamma|)/(1 - |\Gamma|) \tag{1.16}$$

If SWR = 1, the reflection |Γ| = 0, and there is no reflection. The waveguide outlet impedance is fully matched to the inlet (waveguide) impedance, and the propagating power is absorbed entirely in the load. The reflection coefficient can be used to calculate the return loss of the power $P_r$ in decibels (dB):

$$P_r[\mathrm{dB}] = -10 \log|\Gamma| \tag{1.17}$$

If there is an incident power ($P_{inc}$) forwarded by the microwave source and the power reflected from the load is $P_{ref}$, SWR can be expressed as the following:

$$\mathrm{SWR} = \left[1 + (P_{ref}/P_{inc})^{1/2}\right] / \left[1 - (P_{ref}/P_{inc})^{1/2}\right] \tag{1.18}$$

### 1.3.4 Coaxial Transmission Lines

The coaxial transmission lines are often used in connection with the waveguides. The TEM mode propagating in the coaxials has no cut-off frequency and can lead the DC current and voltage. However, typical use of the coaxial lines (cables) is under 3 GHz to avoid high power losses. Typical construction of the coaxial cable is shown in Figure 1.6.

The power level that can be transported through the coaxial cable is dependent on the diameters of the conductors and on the used dielectric

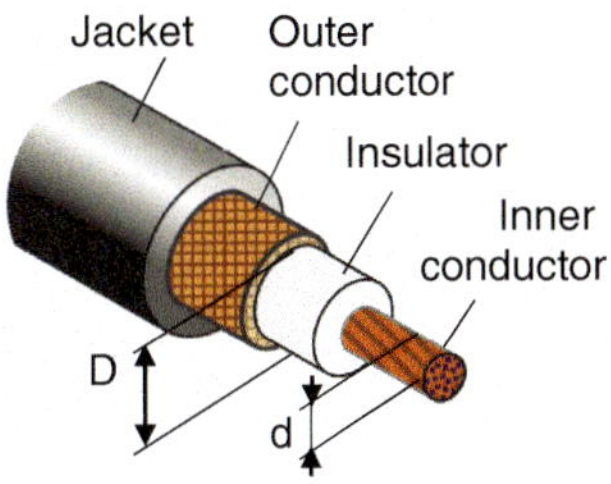


**Figure 1.6** Typical construction of coaxial cables. Symbols D and d are diameters of outer (wrapping the insulator) and inner conductor, respectively.

material. The larger diameters can be obviously used for higher power, but usually not over 2 kW. At higher microwave powers cables can be heated and the dielectric material of the insulator can be melted with consequent short circuits. Even PTFE [$(C_2F_4)_n$] insulators (Teflon) with a melting point of 327°C can cause problems at higher power. The heating can form some carbon parts, which act as the conductive shorts for electric breakdown. The advantage of coaxial cables is their flexibility and possibility of bending. However, if these features are unnecessary and the coaxial can be linear, it is recommended to build water-cooled coaxial lines with air as a dielectric. The construction of such components are presented below.

The impedance of coaxial line can be calculated using the following formula:

$$Z_{coax}\left[\text{Ohm}\right] = 138(\mu_r/\varepsilon_r)^{1/2}\log_{10}\left(D/d\right) \tag{1.19}$$

Typical impedances of commercial coaxial cables are 50 Ohms or 75 Ohms. Expression (1.19) can be simplified assuming that most dielectrics have $\mu_r = 1$, and one can use a natural logarithm (ln or $\log_e$) instead of the decadic one:

$$Z_{coax}\left[\text{Ohm}\right] = 60\varepsilon_r^{-1/2}\ln\left(D/d\right) \tag{1.20}$$

For $Z_{coax} = 50$ Ohms, the required ratio of diameters D/d will be the following:

$$D/d \approx \exp(0.83\varepsilon_r^{1/2}) \tag{1.21}$$

For air ($\varepsilon_r = 1$), it will give D/d ≈ 2.29, and for Teflon ($\varepsilon_r = 2.03$), it will give D/d ≈ 3.26. Such kinds of data allow construction designs of straight coaxial lines with air dielectric and local solid dielectric supports, as shown in Figure 1.7. Because all solid dielectrics have $\varepsilon_r > 1$, the ratio $D_s/d_s$ (diel.) > D/d (air). It can be easier to make $d_s < d$ of the central conductor than larger $D_s > D$ for the outer conductor in the area with the dielectric as shown in Figure 1.7.

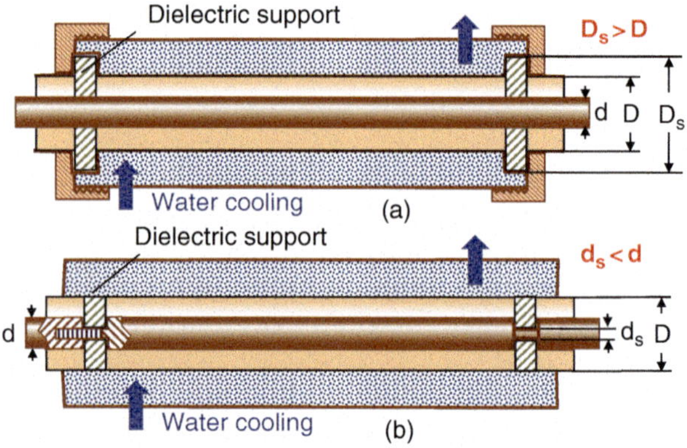


**Figure 1.7** Principle of construction of an air-filled coaxial line with possible cooling for higher powers arranged with dielectric supports of central conductor. (a) The larger diameter $D_s$ of the dielectric support. (b) A shorter diameter $d_s$ of the central conductor.

### 1.3.5 Microwave Resonators

In general terms, any hollow cavity made from conducting material can resonate at certain wavelengths or frequencies of the electromagnetic waves. A simple condition for the resonance in such cavity is that the wavelength of the involved electromagnetic wave ($\lambda$) and the linear dimension of the cavity (length, diameter) $\wedge$ are related to each other:

$$\Lambda = \mathrm{k}\lambda / 2 \; for \; \mathrm{k} = 1, 2, 3, \ldots \tag{1.22}$$

Therefore, using a piece of a waveguide as a resonant cavity is possible, providing both ends are closed by conductive walls and the microwaves are fed into such cavity by an antenna or a narrow slot. Schematic illustrations of typical couplings are shown in Figure 1.8.

The position of the coupling in the waveguide resonator is not arbitrary because the resonances have forms of standing waves with nodes at the conductive ends at both sides. According to Figure 1.5, the maximum wave amplitude in the $TE_{10}$ mode is located at several $\lambda_g/4$ from the node, and there is an optimal place for inserting the power.

## 1.4 Waveguide Power Lines

For the power feeding to any applicator generating the gas discharge plasma, some kind of the waveguide power line must be assembled. Almost all waveguide components and coaxial cables are today commercially available for

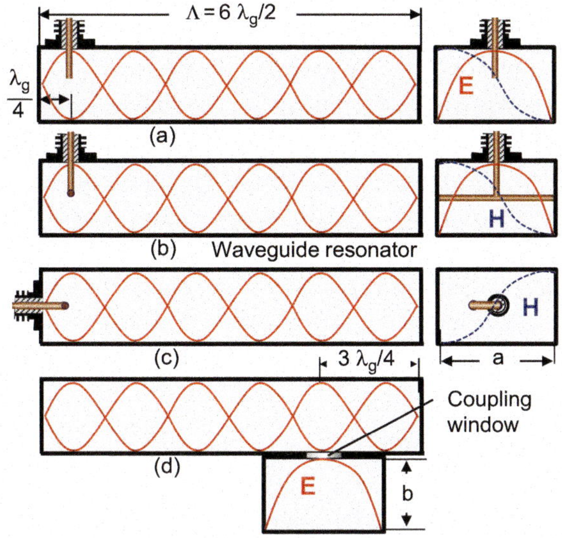


**Figure 1.8** Examples of coupling arrangements. (a) A stub coupling with coaxial. (b) A tee coupling with coaxial. (c) A loop coupling. (d) A slot window coupling with a waveguide.

different frequency ranges and powers. Basic components and their functions will be described below with simple illustrations. Typical power line is composed from or contains parts, shown schematically in Figure 1.9:

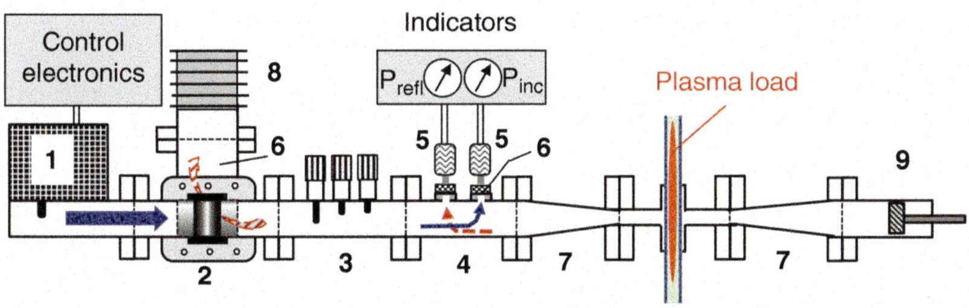


**Figure 1.9** Typical microwave power line for generation of a discharge plasma slab in a dielectric (quartz) tube passing the broader side in a narrowed waveguide.

The numbers in Figure 1.9 mean the following:

1) Microwave generator (usually magnetron) with the waveguide power outlet
2) Insulator (mostly a circulator) protecting the generator from the reflected power
3) Impedance tuner or tuners for maximizing the matching with the load (plasma)

4) Directional coupler or couplers for separating the incident and reflected power
5) Power detectors, e.g. thermistors for measurements of the power
6) Passive waveguide components (bends, coaxial to waveguide couplers, dividers)
7) Tapered waveguide parts or rectangular-to-circular waveguide parts
8) Power loads and load tuners
9) Waveguide shorting plungers

The microwave plasma can be generated inside different shapes of resonators. Typical resonant cavities for the resonant generation of plasmas will be described in Section 1.4.10.

### 1.4.1 Magnetron Tube Microwave Generator

The principles of the magnetron tube operations have been described in Section 1.2. In the laboratory applications of the plasma, the magnetron power is 0.5–5 kW. The generator consists of an electric supply containing a high-voltage feeder, filament heating current circuit, and control electronics. An example of an electric supply for 1.2 kW continual wave (CW) mode magnetron can be found in Ref. [1.9]. The magnetron tube can work in CW mode or in the pulse modes, see Ref. [1.10]. These possibilities should be available from the control electronics. More details will be introduced in Section 1.5. An illustrating example of the design of the magnetron generator with 2.46 GHz 850 W CW magnetron tube, relevant electric circuitry, and the coupling to the $TE_{10}$ waveguide is in Ref. [1.11]. Today, many commercial magnetron power sources are available. Schematic sketch of typical microwave magnetron generator is shown in Figure 1.10.

### 1.4.2 Microwave Insulators

Microwave insulators protect power generators from possible damage by reflected microwaves. The most frequently used systems are called ferrite circulators, which utilize turning the reflected wave by the magnetized ferrite toward a side waveguide arm terminated, e.g. by a dummy load. A simple sketch of such a waveguide circulator is shown in Figure 1.11. Ferrites are anisotropic materials with properties depending on the intensity of applied magnetic field and on the frequency of incident electromagnetic waves. The circulators are always designed for the requested frequency band. At higher powers, the ferrites can be heated during operation, which requires proper cooling, mostly by water.

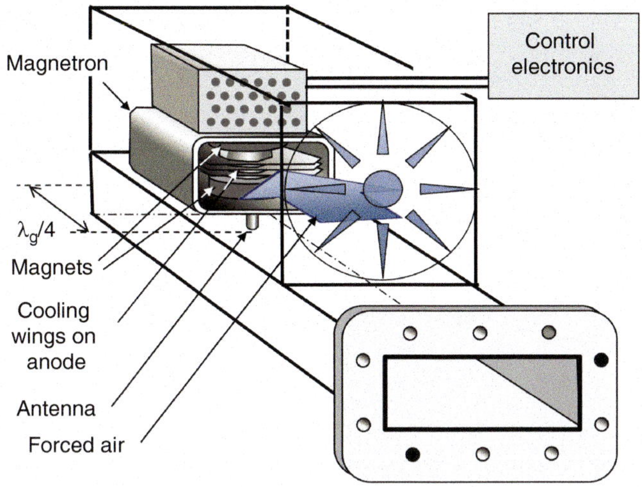


**Figure 1.10** Schematic sketch of typical microwave power generator cooled by the forced air.

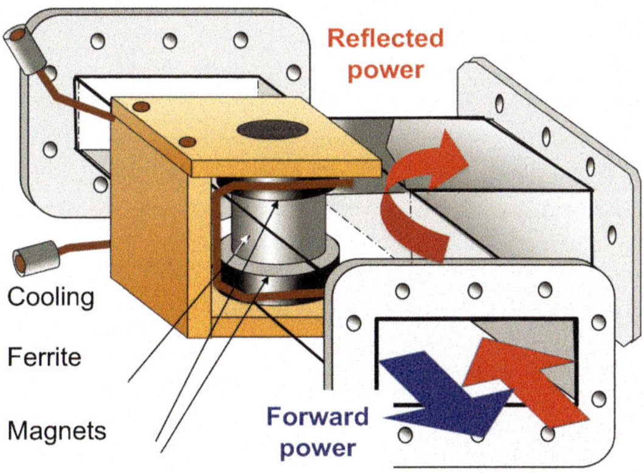


**Figure 1.11** Sketch of the ferrite circulator based on the wave deflection. Directions of the forwarded microwave power and reflected power are depicted by arrows.

### 1.4.3 Impedance Tuners

The incident microwave power from the source should be transferred to the load, e.g. to the gas discharge plasma, at the highest level with minimum reflection. Therefore, tuning the impedance of the waveguide with the plasma load is desirable. Among several systems the most frequently used are three-stub tuners and the E-H tuners (described below).

Any conductor rod immersed in the waveguide represents an impedance disturbance. It can be active, like an antenna, to radiate power from the magnetron generator, or it can have a “passive” function, such as a local shunt susceptance for tuning of the waveguide impedance. A typical representative of a stub-based impedance tuner is arranged with three rods, with stubs having adjustable variable lengths. The stubs are installed in line on the broader dimension of the rectangular waveguide. They can be positioned at distances of $3\lambda_g/8$ from each other. The control of the stub lengths is mostly manual, but electronically driven systems also move stubs to minimize the reflected power. A three-stub tuner is shown in Figure 1.12. An illustration of the E-H tuner is in Figure 1.13. The E-H tuner has two waveguide "arms" for affecting the E-field and H-field in the main waveguide. The principle is based on two arms forming parallel impedances tuned by shorting plungers, adjusted manually or by motors.

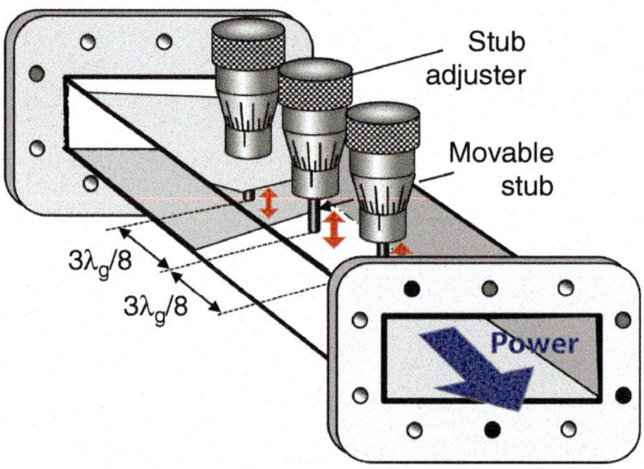


**Figure 1.12** Schematic illustration of the three-stub tuner installed in the rectangular waveguide.

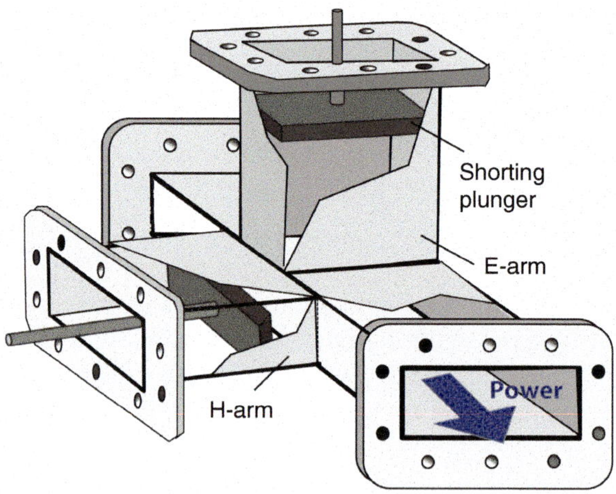


**Figure 1.13** Schematic sketch of the E-H tuner. Both arms are equipped by shorting plungers.

### 1.4.4 Directional Couplers

Directional couplers can sample certain amounts of the microwave power in the selected direction (forward and/or reflected). This ability is used for measurements of the forward and reflected power, necessary for tuning the impedance and minimizing the reflections. In Section 1.3.3, it was explained that the SWR can be used for measurements of the reflected power. In a simple approach, two points at distance $\lambda_g/4$ should always coincide with higher and lower signals (up to the maximum and minimum) in the SWR or VSWR. Proper placement and shape of the slots made in the waveguide wall can be used for the construction of systems measuring part of the incident power (or voltage in the VSWR) and the relevant reflections. Many constructions of directional couplers exist with waveguide arms installed on the E-side or at the H-side of the main waveguide. Older types of couplers typically used an extra waveguide arm with coaxial connector at one end and a dummy load on the opposite end. They were installed on the broader wall of the waveguide and coupled to the main waveguide by at least one pair of the connecting long slots. Advanced loop directional couplers use conductive strips connected to the load resistors. Two of the loops are installed in two circular cavities mounted on a broader wall of straight waveguide piece with circular coupling holes. An illustrative sketch of such a coupler is in Figure 1.14.

As can be seen, each cavity is coupled with the main waveguide by a small circular hole. The conductive loop strip coupled to the coaxial connector on the cavity serves for measurement of the signal, and an opposite resistor represents a dummy load, see, e.g. Ref. [1.12]. Knowing the coupling of the cavity with the waveguide, calibrating such a system and measuring the power in both directions is possible.

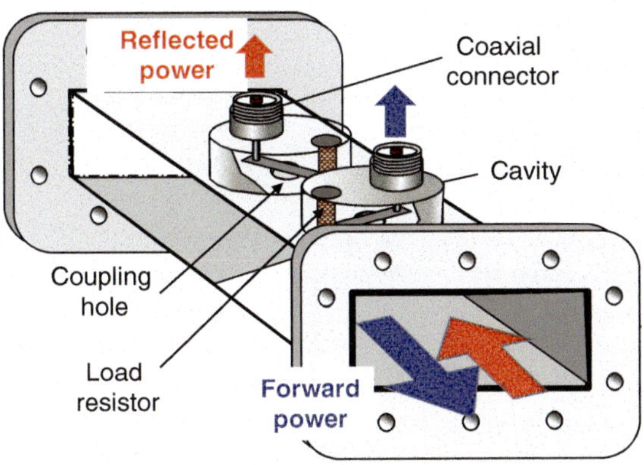


**Figure 1.14** Loop-type double directional coupler. Coaxial connectors serve for coupling to power detectors.

The coaxial outputs of the microwave coupler are usually connected to the indicator heads, which can be equipped by thermal absorbing sensors (called bolometers) or fast diodes. The thermal sensors, thermistors in the bolometer heads, are semiconductors with the strongly dependent conductivity on the absorbed heat. The microwave diodes are VSWR elements fast responding on the measured voltage changes. Many commercial systems are available for such microwave power measurements.

### 1.4.5 Passive Waveguide Components – Bends, Flanges, Vacuum Windows

Rectangular waveguides often require bended components to fit the power line into a limited space. There are several types of bends having different bending angles and constructions. Individual constructions differ in the wave reflections. The best bends with minimum reflections can be formed by smooth bending with the middle length of about $2\lambda_g$. Illustrations of such 90° bends in the E direction and the H direction are shown in Figure 1.15.

In all illustrations of the rectangular waveguide components showed in Figures 1.10–1.15, the connecting flanges were flat and provided with 10 holes for screwing together. However, in many practical cases, particularly at high powers, such simple connections can cause strong reflections and visible sparking between flanges. To improve waveguide connections, the pair of the contact flanges can be designed with a quarter wavelength choke slit working as a resonator, see Figure 1.16. In such a connection, the choke has high input impedance, which prevents waves to move into the slit and leads to negligible coupling impedance at the broader wall ("a") of the contacting waveguides, where $TE_{10}$ mode has maximum intensity of the electric field E. The choke

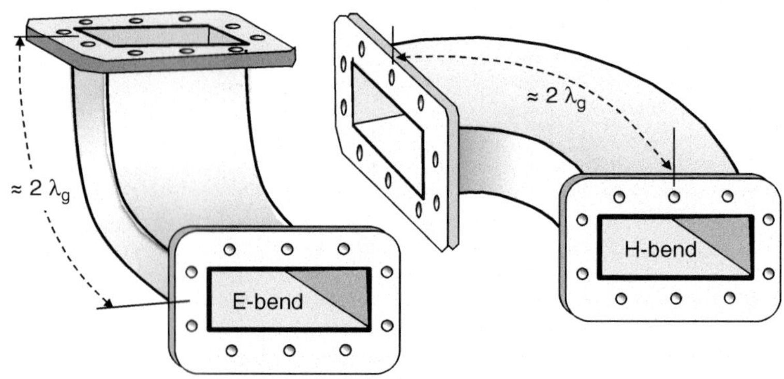


**Figure 1.15** Illustrations of typical 90° waveguide bends in the E direction and H direction.

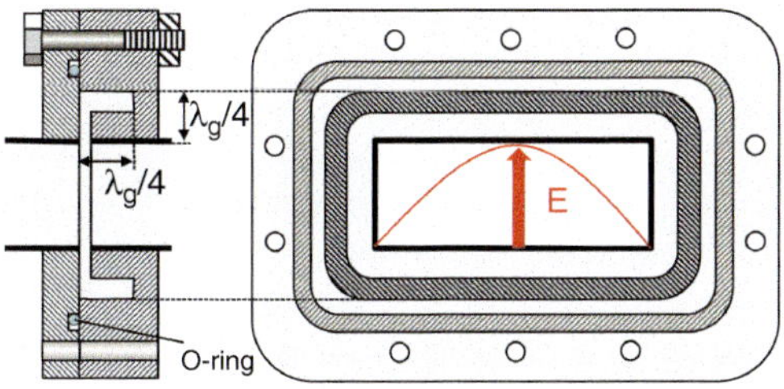


**Figure 1.16** An illustration of a choke-type flange connection between waveguide components.

groves are made only on one flange, and the opposite flange is flat to close the choke arrangement. When required, the choke coupling in Figure 1.15 can contain a groove for an O-ring sealing the waveguide system for the pressurizing or evacuation.

The rubber of the O-rings used for the sealing of waveguide parts shouldn't be from "ordinary" rubber (e.g. Viton, Neoprene, or Nitrile) because of possible damage by the microwave absorption and heating/burning damage the O-ring as shown in Figure 1.17. To avoid problems with rubber damage, use O-rings made of the silicone rubber. The silicone rubber withstands higher temperatures (up to about 300°C) and is less sensitive to the microwave heating based on the power absorption. Because of silicone rubber's good electric insulation ability, it is often used to prevent arcing on insulators in the high-voltage power lines.

An often requirement is to transport the microwave power from air filled line to an evacuated or pressurized waveguide part. Practical construction of suitable window used for the power level up to 2 KW is shown in Figure 1.18.

**Figure 1.17** Microwave caused damage of an ordinary Viton O-ring used for vacuum sealing.

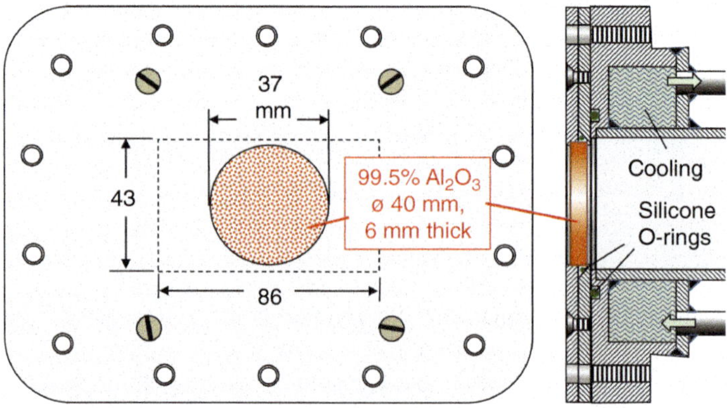


**Figure 1.18** Construction of a sealed ceramic window for $TE_{10}$ waveguide 2.4 GHz. The back flange on the vacuum side is water cooled. The flange of the coupled waveguide (not shown) can be flat or a choke type.

The pair of flanges holding the alumina window shown in the figure, can be made from 4 mm thick aluminum plates. The window is sealed by silicone O-ring having diameters 40 x 37 x 2 mm placed in a conical saddle groove pressing the O-ring.

### 1.4.6 Tapered Waveguides and Waveguide Transformers

As explained above, the $TE_{10}$ dominant mode in the rectangular waveguides has a maximum electric field intensity (E) in the center of the broader waveguide wall at several quarter wavelengths distances from a shorting wall (plunger) reflecting the wave. Because the electric field is defined as voltage over distance (in this case, the height "b" of the waveguide), i.e. E = V/b, the decreasing dimension b should increase the intensity of the electric field. This is suitable, e.g. for an easier generation of a plasma slab in a dielectric tube passing the waveguide with the maximum E, as in power line shown in Figure 1.9. Decreasing the dimensions can be made in the shorter ("b") but also in the wider ("a") wall, or even in both waveguide cross dimensions. The components with shortened cross-dimensions providing minimum reflections are called *waveguide transformers*. If the input impedance of such component is $Z_{in}$ and the output one should be $Z_{out}$, the transformation can be provided by a quarter wave section with an impedance $Z_{tr}$:

$$Z_{tr} = \left(Z_{in} \cdot Z_{out}\right)^{1/2} \tag{1.23}$$

In most cases, the impedance transformation is made in the E direction (dimension b) to increase the electric field E. It can be provided by smoothly

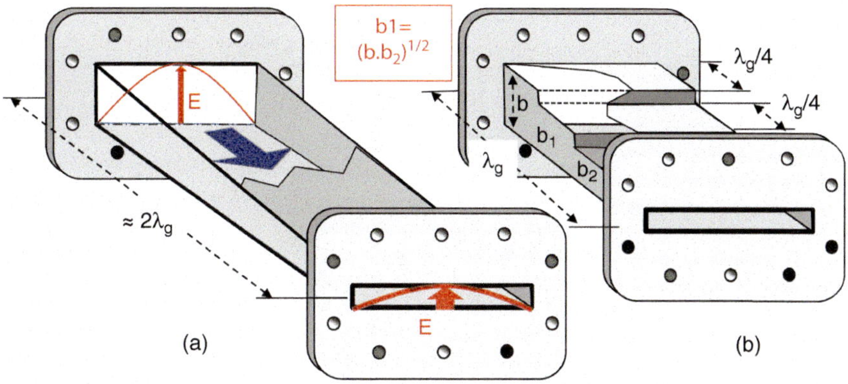


**Figure 1.19** (a) Tapered waveguide. (b) Four sections of a quarter wave transformer (the last section is not seen).

tapering a waveguide over the length of about 2 $\lambda_g$, see Figure 1.19a. The quarter wave transformer is shown in Figure 1.19b.

For the shortest waveguide transformer the minimum number of the transforming steps is one, where the dimension $b_1$ is the following:

$$b_1 = \left(b.\ b_2\right)^{1/2} \tag{1.24}$$

However, the step transformer can have more quarter wave steps by applying the similar rule for the intermediate dimension b as given in Eq. (1.24).

In the case of transforming the wave by shortening the broader dimension "a" of the waveguide, it changes the waveguide wavelength $\lambda_g$ in the individual sections. The relevant calculation is more complicated, see Ref. [1.13]. In a "one-step" transformer from the basic waveguide dimension "a" and wavelength $\lambda_g$ to the dimension "$a_2$" with wavelength $\lambda_{g2}$, the intermediate section will have dimension $a_1$ and wavelength $\lambda_{g1}$, where the equation applies:

$$a_1/\lambda_{g1} = (a/\lambda_g)^{1/2}(a_2\,/\,\lambda_{g2})^{1/2} \tag{1.25}$$

### 1.4.7 Power Loads and Load Tuners

In many practical cases, entirely attenuating or absorbing the microwave power is necessary in a dummy load. Such attenuators can be designed as waveguide sections with the cut-off condition, where the operating frequency is far below the cut-off frequency of the incident wave or the wavelength is above the cut-off wavelength, see Eqs. (1.2, 1.3). The wave is unable to propagate in the waveguide. As the cut-off wavelength $\lambda_{coff}$ for dominant mode $TE_{10}$ is 2a, see Eq. (1.3), narrowing the dimension "a" in the waveguide is possible to

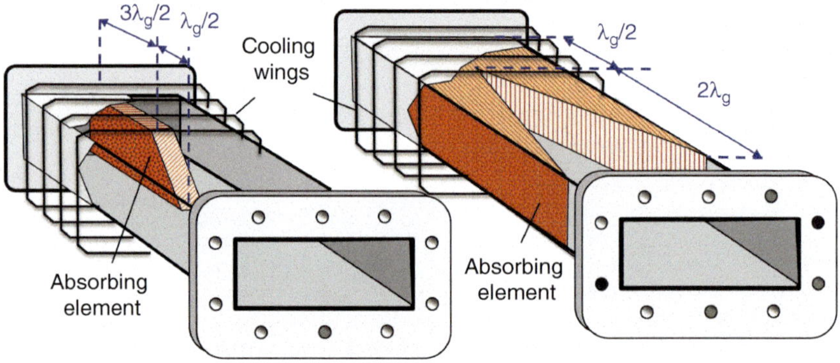


**Figure 1.20** Examples of waveguide attenuators. Air cooling wings can be replaced by a barrel with cooling water.

settle the cut-off condition. However, in other arrangements, a piece of absorbing material (e.g. glass) is installed in the waveguide and shaped with a sharp end to decrease reflections, see examples in Figure 1.20. Some details can be found, see e.g. Ref. [1.14].

The absorption of the electromagnetic power at microwave frequencies depends on properties of dielectric material used as a load, and the microwave power is transferred to the heat. A simple example can be the microwave oven with the power absorption in water contained in the food, see Section 1.5. However, other materials exist with high absorption coefficient. The measure of the losses by the wave absorption is called the loss tangent (tan $\delta$). The angle $\delta$ in any dielectric material with the complex relative permittivity $\varepsilon_r$ represents the angle between the real part of the complex permittivity (resistive part) and its reactive part. The lower tan $\delta$ is, the lower the dielectric losses are. For an ideal value of $\tan \delta = 0$, it means $\delta = 0$, and that means no loss. A comparison of the loss tangent values for selected materials measured between 1–8 GHz is shown in Table 1.1.

The loss tangent depends on the wave frequency and the temperature of material. Data about dielectric losses of different materials can be found in Ref. [1.15].

An outstanding material for the microwave absorption and heating applications is silicon carbide (SiC), e.g. made in the form of Carborundum. Graphite is also an efficient absorbing material when used in form of fibers, see Ref. [1.16]. For high-power absorption loads, using ordinary water-cooling using glass pipes is useful, see Figure 1.21. For the rectangular waveguide ($43 \times 86$ mm$^2$) the suitable glass pipe with outer diameter 22 mm and inner diameter 18 mm should be from the Pyrex or quartz glass. The absorption principle is similar as in the microwave ovens.

**Table 1.1** Comparison of the loss tangent values of selected materials from 1–8 GHz.

| Material | tan δ | Material | tan δ |
|---|---|---|---|
| Alumina | $2 \times 10^{-4}$ | Neoprene rubber | $3.7 \times 10^{-2}$ |
| Fused quartz | $6 \times 10^{-5}$ | Silicone rubber | $1 \times 10^{-3}$ |
| Corning glass | $3.6 \times 10^{-3}$ | Polyimide | $6 \times 10^{-3}$ |
| Teflon | $2.8 \times 10^{-4}$ | Water | $1.6 \times 10^{-1}$ |
| Silicon | $1.7 \times 10^{-3}$ | Graphite fibers | $1 \times 10^{-2}$ |
| Silicon carbide | $3 \times 10^{-3}$ | Carborundum | $5.8 \times 10^{-1}$ |

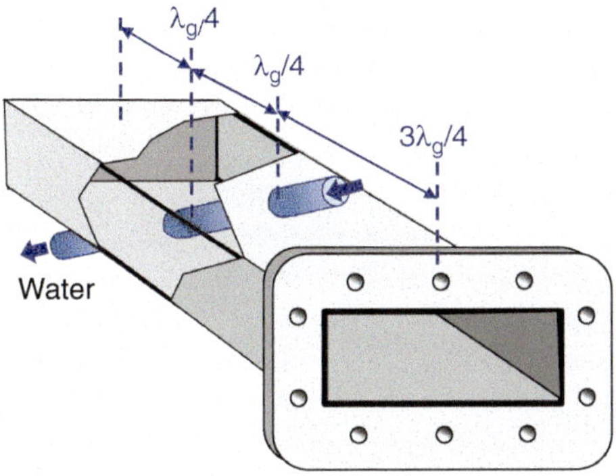


**Figure 1.21** An illustration of the microwave load with the water tube for absorbing the power.

### 1.4.8 Waveguide Phase Shifters

For tuning, e.g. electric field E, in any local part of the waveguide, adjusting the phase of the wave is possible by inserting some elements made from low-loss dielectric into the waveguide, which can change the effective dielectric constant in this part and the wave phase without changing the wavelength. Even with the low-loss dielectric, the above proposed option is limited for low powers due to possible heating damage of the dielectric. Tunable phase shifters have different constructions, see, e.g. Refs. [1.17, 1.18]. The position of the dielectric element in simple waveguide phase shifters can be changed manually and by motor shifting controlled by a control unit. A simple phase shifter with a movable dielectric "paddle" is shown in Figure 1.22.

### 1.4.9 Waveguide Shorting Plungers

In many installations of the microwave power lines, the waveguide should be terminated by some kind of movable shorts, see, e.g. Ref. [1.19]. The simplest shorts use movable shorting plungers, see Figure 1.23. The plunger in Figure 1.23a has ohmic contact due to sliding of a metal "comb" with the waveguide

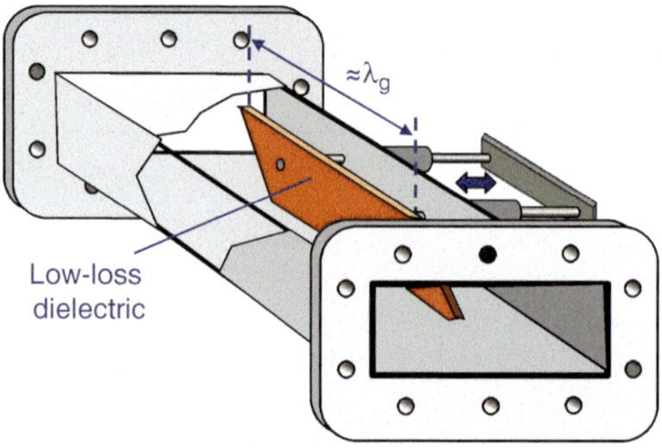


**Figure 1.22** An example of the waveguide phase shifter with movable dielectric "paddle".

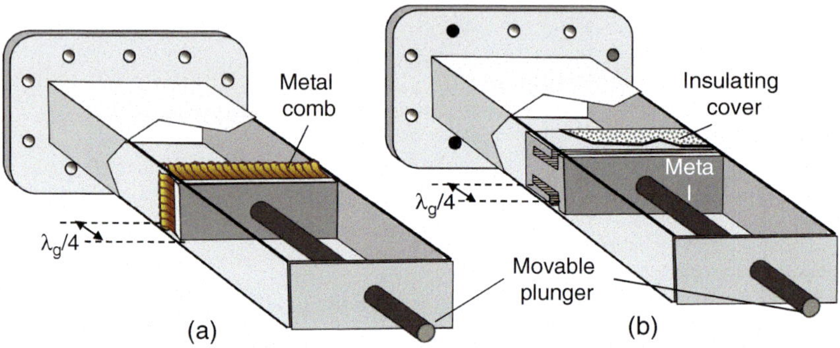


**Figure 1.23** Shorting plungers for rectangular waveguides. The contacting plunger with sliding metal comb (a). A non-contacting plunger with a single choke made in the metal part (b).

walls. The choke-based plungers have no ohmic contact with the waveguide, see Figure 1.23b. Such a plunger can be made from a good conductive bulk metal, e.g. aluminum, packed in an insulating low loss dielectric cover.

### 1.4.10 Coupling from Rectangular to Circular Waveguide: Resonant Cavities for Generation of Plasma

As shown in Figure 1.8 (Section 1.3.5), the microwave resonator can be made as a piece of rectangular waveguide having length $\wedge = k\ \lambda/2$ for k = 1, 2, 3,.... However, in different systems for plasma generation, the resonators can have other more suitable shapes. The most obvious is a cylindrically shaped metal chamber, where the plasma is generated in an inserted quartz cylinder or bell jar chamber. There are three types of connections between the rectangular waveguide power line and the cylindrical waveguide used as a resonator:

1) A smooth $\geq 2\lambda_g$ with a long transition between the rectangular waveguide with $TE_{10}$ mode and the circular one with $TE_{11}$ mode (see, e.g. Ref. [1.20])

2) An antenna-based coupling
3) An orifice hole coupling between waveguides

The smooth transition coupling between rectangular and cylindrical waveguide is shown in Figure 1.24.

The arrangement in Figure 1.24 can be used for couplings of a rectangular waveguide with a circular resonator either directly, or e.g. via an orifice between them. Somewhat simpler coupling arrangements can be performed with an antenna and an orifice or by some window, iris, or slit transitions. The shape and positions of the coupling orifices affect the resonant mode excited in the cavity. Cylindrical resonant cavities are often made from circular waveguides by putting conductive side walls. As the dominant mode in the circular waveguide is $TE_{11}$ (see Figure 1.3), the dominant mode for the cylindrical resonant cavity with the length L longer than diameter 2r is typically $TE_{111}$. However, in short resonators, where the length is $\wedge < 2r$, the dominant mode would be $TM_{010}$. The geometry of the electric and magnetic fields in the resonator $TE_{111}$ is shown in Figure 1.25. The resonance frequency of such a resonator filled by air or evacuated (i.e. where relative permittivity and permeability are equal to one) can be expressed as the following:

$$f_{red} = \left[1.841^2 + (\pi r / \wedge)^2\right]^{1/2} / 2\pi r \tag{1.26}$$

This expression is also shown in Figure 1.25. Related theoretical backgrounds can be found in Refs. [1.20, 1.21].

More detailed descriptions of the cylindrical waveguide resonators and their couplings to the incident power are described, e.g. in Refs. [1.22–1.24]. Theoretical background can be found in Chapter 6 of the Ref. [1.21]. Figure 1.26 shows two examples of resonator couplings to the antenna and to an orifice. The used antenna can be made with tunable length, which can optimize the exciting efficacy and lowering of the reflections. The cavity can be arranged with a movable plunger on one side, which can fine-tune the resonances.

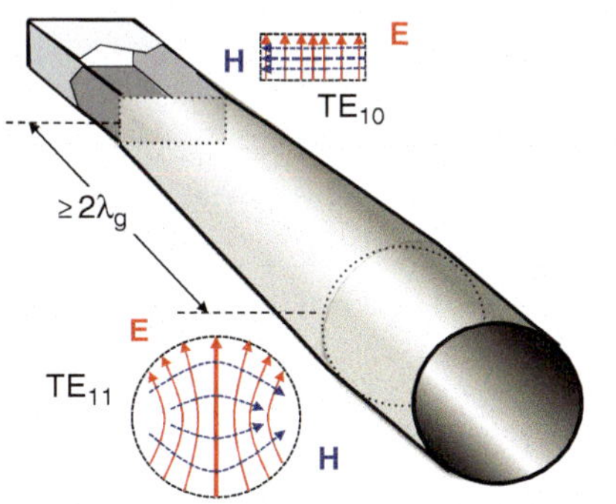


**Figure 1.24** The simple smooth transition between a rectangular and a circular waveguide. The diameter of the circular waveguide should exceed the diagonal dimension of 96 mm of the rectangular waveguide.

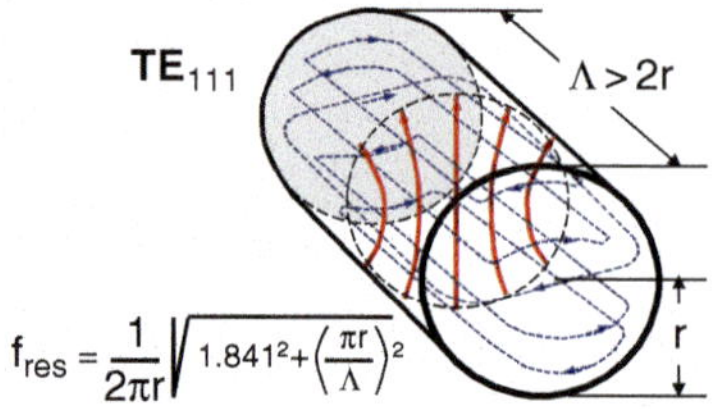


**Figure 1.25** Distributions of the electric and magnetic fields in the cylindrical waveguide resonator with dominant mode $TE_{111}$. The resonant frequency is given for an empty cylinder.

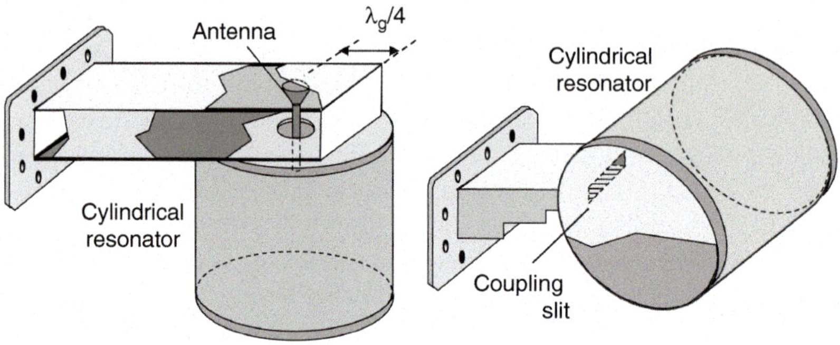


**Figure 1.26** Examples of the coupling between a resonator and a rectangular waveguide. The coupling by an antenna (a). Coupling by an orifice (b).

The coupling orifices can have different shapes and sizes for optimal power coupling. Rectangular orifices can have an angle with respect to the axis of the resonator. Moreover, the waveguide can be coupled to the resonator also tangentially. All these options can affect the coupling efficiency and the modes excited in the resonator.

## 1.5 Microwave Oven – A Most Common Microwave Power Device

The microwave ovens are common devices in many households, kitchens, restaurants, rest areas in companies, and refreshment rooms. Figure 1.27 shows a typical student refreshment area at university with multiple ovens for heating food quickly. However, every oven represents a smartly engineered microwave system, which can help us understand the basic principles and components in the microwave technology introduced and described above.

From Table 1.1 (in Section 1.4.7), it is understood that heating food is based on the efficient absorption of the power at frequency of 2.4 ± 0.05 GHz in the water contained in the food. The human body is composed of more than 60% water and, therefore, exposing any body part to the power generated by the microwave oven is dangerous. Of particular danger are the human eyes, which

**Figure 1.27** Typical service arrangement with multiple microwave ovens for university students.

can be damaged with a short exposure, which is why all microwave ovens have their door windows protected by the shielding metal net. A similar warning extends to animals like dogs or cats. The microwave radiation can be used also for killing of unwanted insects, e.g. with the storage of flour.

The ovens are designed to distribute the microwaves as uniformly as possible over the food. At the same time, the system is constructed to be as simple and cheap as possible. The components used in each oven are the magnetron tube (usually for 2.45 GHz) described in Sections 1.2 and 1.4.1 and a piece of the rectangular waveguide for the $TE_{10}$ modes (see Section 1.3), with an open end radiating the microwaves into the main chamber of the oven. The chambers are made from the metal sheets to reflect the waves, and the walls are covered by dielectric sheaths for easy cleaning. Therefore, the chamber represents a multimode resonant cavity. To prevent the waveguide from grease and fat vapor, the radiating waveguide outlet is covered by a special dielectric plate made usually from mica. After long use, this part can show enhanced dielectric heat damage or some sparking, and it needs to be replaced.

Many commercial magnetrons are available for microwave ovens. In the ovens for domestic use, these tubes are air cooled and give maximum power of about 0.5–1 kW. Figure 1.28 shows a typical magnetron tube, e.g. WB27X10516 OM75P(31) for GE Samsung. Most of these microwave generator tubes are relatively cheap (often below US$ 100). In domestic ovens, the antenna delivering the power is immersed into the waveguide section, as illustrated in Figure 1.9 in Section 1.4.1.

The power can be radiated into the cooking cavity from the top, the side, or from the bottom, and the cavity is a multimode resonator. The most typical arrangements are from the top or side (see Figure 1.29). To distribute the microwave

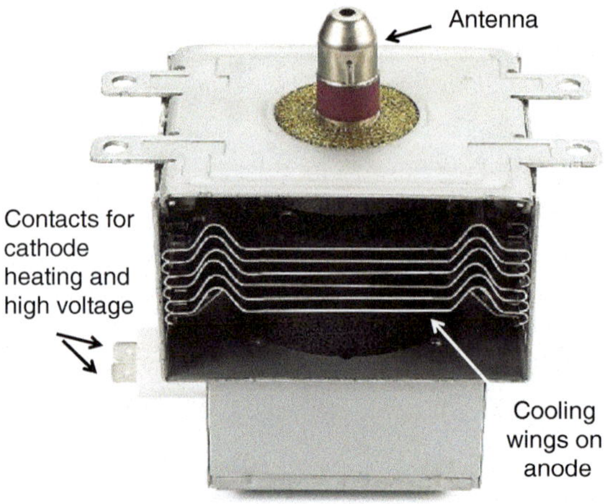


**Figure 1.28** Shapes of typical air-cooled microwave magnetron tubes used in microwave ovens.

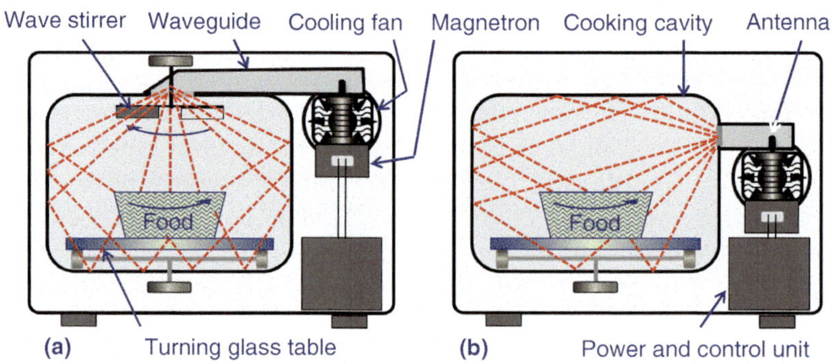


**Figure 1.29** Schematic illustrations of heating arrangements in the microwave ovens. (a) The top arranged waveguide with the wave stirrer, (b) the side arranged waveguide.

heating power, the food is placed on a turning glass holder and the power outlet in the cooking cavity is often arranged with the rotating wave stirrer reflecting the waves to different angles.

Figure 1.30 shows a typical view into the microwave oven. This oven is equipped with a control knob for adjusting the heating time on a digital indicator, as recommended for different kinds of the food. The knob can be used for adjusting different heating regimes. As illustrated in Figure 1.29, the food should be placed on a rotating glass table for optimization of the heating action.

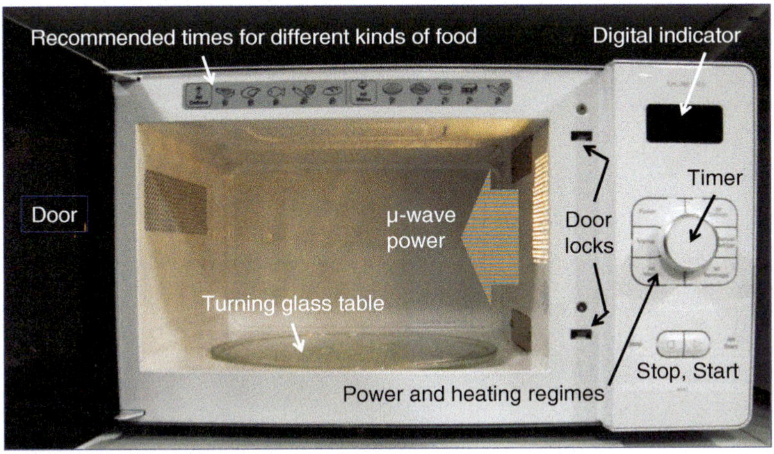


**Figure 1.30** A view into a typical microwave oven with the power outlet arranged on the right side.

The domestic microwave ovens are constructed for simple heating, and they use magnetron tubes with the power ≤ 1 kW. In such magnetrons, the typical values for the cathode filament heater are about 2 V and 10 A AC, and the rectified high DC voltage at the cathode is about 3–5 kV. As mentioned in Section 1.2, the amplitude of the DC voltage must exceed a threshold to initiate the current between the cathode and anode. Because no special filtration of the DC voltage is used in the ovens, the high voltage is pulsing with the frequency of the electricity (50 or 60 Hz) as does the output microwave power. This is acceptable for the heating of food but not for microwave systems intended for generation of stable plasmas. Moreover, for the generation of plasmas, the power lines are usually required with good control of the output power.

Figure 1.31 shows the basic electric circuit for the powering of the magnetron tubes in the microwave ovens. A simple circuit is presented here without components controlling the power level, possible pulsing regimes, time, overheating sensors, etc. The high DC voltage for the magnetron is provided by doubling and rectifying the high AC voltage from the transformer using a high-voltage capacitor and a high-voltage diode. Because this high DC voltage remains on the capacitor even after switching off the power, any unskilled manipulation inside the ovens is dangerous, and it could bring serious life risks.

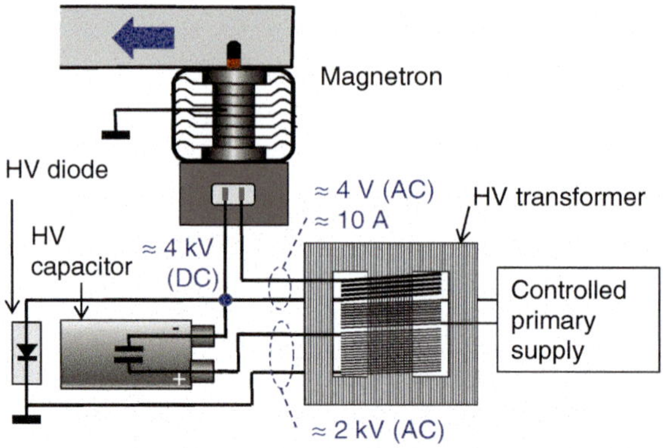


**Figure 1.31** An illustration of simple circuitry used with magnetron tubes in microwave ovens.

The magnetron power depends on the cathode filament current controlling the emission of electrons from the cathode and on the level of the high DC voltage controlling the DC current to the anode. The power control is complex because the DC current changes rapidly with the cathode voltage above the threshold. Keeping the stable level usually requires current limiting arrangements in the transformer. The variation of the heater current is an option, but it needs a corresponding control circuit. The magnetron operation in an on-off regime needs proper switches, and it reduces the magnetron lifetime. Such options are arranged in more sophisticated commercial microwave power sources, whereas household ovens mostly utilize only constant power (with some oscillations) and the heating is controlled by the operation time.

The microwave ovens are equipped with neither ferrite insulators protecting the magnetron tube from the reflected power nor with impedance tuning systems and indicators, as required in the microwave power lines for plasma experiments and applications. Despite its simplicity, the microwave oven is a microwave power source and can be used at certain conditions for the ionization of a gas and generation of a gas discharge plasma. Many attempts have been tried to use the microwave oven as the plasma source, see, e.g. Ref. [1.25].

An introduction to the plasma and plasma-related processes will be given in the next Chapter 2. However, at this stage, we can illustrate the possibility of igniting the plasma in an ordinary microwave oven. Figure 1.32 presents a simple experiment.

In Figure 1.32, we have used a quartz tube with a mercury vapor at reduced pressure. Such tubes were used in the ultraviolet (UV) disinfection treatments of water. The upper photograph shows the tube placed on the turning glass table inside the oven. The lower photograph shows an ignited Hg discharge during a short switching on of the microwave power.

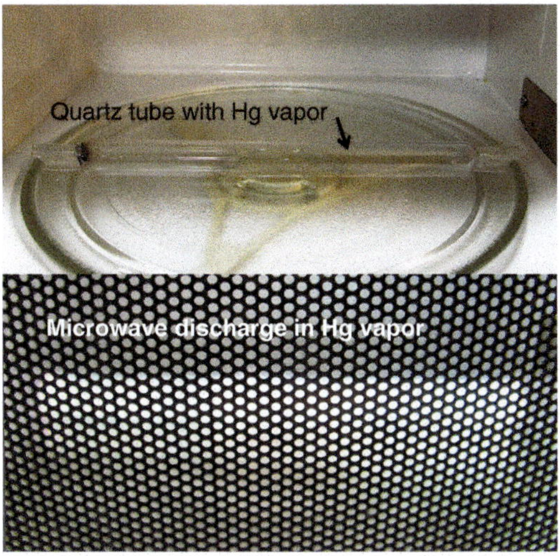


**Figure 1.32** An illustration of the mercury (Hg) plasma generated by the microwave power inside the microwave oven. The discharge is shown behind the net-shielded door of the oven. The intensity of the discharge is not uniform and it depends on the position of the table.

## References

[1.1] J. Al-Khalili: "The birth of the electric machines: a commentary on Faraday (1832) 'Experimental researches in electricity'"; Philos. Trans. Royal Soc. A373 (2015) 20140208, pp. 1–12. http://dx.doi.org/10.1098/rsta.2014.0208.

[1.2] National Aeronautics and Space Administration, Science Mission Directorate: "Anatomy of an electromagnetic wave", 2010. Retrieved [insert date – e.g. August 10, 2016], from NASA Science website: http://science.nasa.gov/ems/02_anatomy.

[1.3] S. Hong: "Wireless: From Marconi's black-box to the audition", The MIT Press Cambridge, Massachusetts, London, England, 2001. ISBN 0-262-08298-5.

[1.4] M. Guarnieri: "The early history of radar"; IEEE Industrial Electronics Magazine, September (2010) 36–42.

[1.5] "The latest word in communications"; "Microwave" New York-Boston System. By Long Lines Department, Plant Division One, New York. November 13, 1947. http://d2xunoxnk3vwmv.cloudfront.net/uploads/1947-ATTs-First-Microwave-Radio-Route.pdf.

[1.6] R.S. Symons: "Modern microwave power sources"; IEEE AESS Systems Magazine, January (2002) 19–26.

[1.7] C.O. Kappe: "Controlled microwave heating in modern organic synthesis"; Angew. Chem. Int. Ed. 43 (2004) 6250–6284. http://dx.doi.org/10.1002/anie.200400655.

[1.8] J. Sun, W. Wang, and Q. Yue: "Review on microwave-matter interaction fundamentals and efficient microwave-associated heating strategies"; Materials 9 (4) (2016) 231, 25 pp. http://dx.doi.org/10.3390/ma9040231.

[1.9] M. Heggannavar and H. Kulkarni: "Power supply design for magnetron power source from single phase supply": 2015 International Conference on Energy Systems and Applications (ICESA 2015), Patil Inst. of Engin. and Technol., Pune, India 30 Oct–01 Nov 2015. https://ieeexplore.ieee.org>document.

[1.10] V.K. Yadav, K. Santhyanarayana, D. Purohit, and D. Bora: "A tetrode based fast pulsed microwave source for electron cyclotron resonance breakdown experiments": Rev. Sci. Instrum. 78 (2007) 023503. http://dx.doi.org/10.1063/1.2671793

[1.11] V. Surducan, E. Surducan, R. Ciupa, and C. Neamtu: "Microwave generator for scientific and medical applications"; Processes in Isotopes and Molecules (PIM 2011), AIP Conf. Proc. 1425, 89–92 (2012). http://dx.doi.org/10.1063/1.3681974.

[1.12] H. Ishibashi, M. Kurihara, Y. Tahara, H. Yukawa, T. Owada, and H. Miyashiya: "Waveguide loop-type directional coupler using a coupling conductor with protuberances"; Proceedings of the 45th European Microwave Conference, 7–10 Sept 2015, Paris, France, 1026–1029. 978-2-87487-039-2 © 2015 EuMA.

[1.13] A. Das and S.K. Das: "Microwave engineering", 3rd Edition, McGraw-Hill Education, India, 2015, Chapter 5.4.3 "Realisation of Waveguide Impedance Transformer", p. 121.

[1.14] P.K. Verma and R. Kumar: "Realization of wide band waveguide terminations at Ku-Band"; Int. J. Adv. Microwave Technol. 3 (4) (2018) 176–179. http://dx.doi.org/10.32452/IJAMT.2018.176179.

[1.15] W.B. Westphal and A. Sils: "Dielectric constant and loss data", Tech. Report AFML-TR-72-39, MIT, April 1972. https://apps.dtic.mil/dtic/tr/fulltext/u2/746686.pdf.

[1.16] J. Fukushima, S. Tsubaki, T. Matsuzawa, K. Kashimura, T. Mitani, T. Namioka, S. Fujii, N. Shinohara, H. Takizawa, and Y. Wada: "Effect of aspect ratio on the permittivity of graphite fiber in microwave heating"; Materials 11 (169) (2018) 1–13. http://dx.doi.org/10.3390/ma11010169.

[1.17] G.J. Halford: "A wide-band waveguide phase-shifter"; Radio and Communication Engineering 100 part III (65) (1953) 117–124.

[1.18] B.T.W. Gillatt, M. D'Auria, W.J. Otter, N.M. Ridler, and C. Lucyszyn: "3-D printed variable phase shifter"; IEEE Microwave Wirel. Compon. Lett. 26 (10) (2016) 822–824.

[1.19] V. Bilik and J. Bezek: "Analysis of modes in rectangular-waveguide noncontacting shorting plunger"; Slovak Univ. Technol., Faculty of Electrical Engineering and Information Technology, Bratislava, Slovakia. Web: https://duepublico2.uni-due.de/servlets/MCRFileNodeServlet/duepublico_derivate_00014694/Final_Papers/GM0073-F.pdf.

[1.20] A. Munir and M.F.Y. Musthofa:: "Rectangular to circular waveguide converter for microwave devices characterization"; Int. J. Electr. Eng. Inf. 3 (3) (2011) 350–359.

[1.21] D.M. Pozar:: "Microwave engineering", 2nd Edition, John Wiley & Sons Inc., New York, 1996.

[1.22] Y. Batygin: "Proton and ion linear accelerators. 2. Introduction to accelerating structures"; U.S. Particle Accelerator School, Albuquerque, New Mexico, June 17–28, 2019. https://uspas.fnal.gov/materials/19NewMexico/PILA/2.Introduction_to_Accelerating_Structures.pdf.

[1.23] R.S. Kashyap: "Waveguide resonators"; Term paper report on waveguide resonators: EE 614: Solid State Microwave Devices & Applications, Dept. of Electric Engineering, Indian Inst. Technol. Bombay. http://citeseerx.ist.psu.edu/viewdoc/download?doi=10.1.1.108.8566&rep=rep1&type=pdf.

[1.24] A. Nassiri: "Wave guides and resonators"; US Particle Accelerator School USPAS, Lecture 5, USPAS 2010, MIT, pp. 1–62. https://uspas.fnal.gov/materials/10MIT/Lecture6.pdf.

[1.25] V.J. Law and D.P. Dowling:: "Converting a microwave oven into a plasma reactor: a review"; Hindawi Int. J. Chem. Eng. 2018, May 25 2018. Article ID 2957194, 12 pages. https://doi.org/10.1155/2018/2957194.

# 2

# Gas Discharge Plasmas

This chapter is devoted to the explanations of fundamental principles important for the understanding of discharge plasmas, particularly those where the plasma is generated by an alternating current (AC) power at high frequencies. The main emphasis is given to the plasmas generated by the microwave power. Sections of the texts and illustrations are taken from the authors' annual courses C-210 and C-324 since 1997 for the Society of Vacuum Coaters (www.svc.org) in the United States and from university courses and many seminars.

## 2.1 Basic Understanding of the Gas Discharge Plasmas

We recognize three basic states of the matter: solid, liquid, and gas. By delivering some kind of energy to a solid, e.g. thermal energy by heating, its temperature T increases, and at the melting temperature, the solid is melted to a liquid. By delivering more heat, the liquid boils and is evaporated to the gas phase. If the heat continues to increase, part of the neutral gas particles loses electrons and become positively charged ions. This last step of the process is called *ionization*. However, due to collisions with other particles, the ions can get the electrons back again and they *recombine* to the neutral particles. Therefore, the charged particles (ions) generated by the ionization have limited lifetimes, and they vanish (recombine) due to interactions with other particles in the volume, mainly with the neutral particles under the assistance of electrons, with electrons on solid surfaces in the reactor vessel, or due to the Coulomb forces between opposite charges of ions and electrons. The collisions lead to the removal of charges, i.e. to a discharging of the gas, which is accompanied by deliberation (radiation) of an excess energy, partly in the form of visible light observed in the discharge chambers, hence, the common name *gas discharge*. At a certain level of the delivered power, the number of ions (ionization) prevails over the ongoing recombination, the gas continues being partly ionized and electrically conductive and remains in the form of a *plasma* or *gas*

*Microwave Plasma Sources and Methods in Processing Technology*, First Edition.
Ladislav Bárdoš and Hana Baránková.

*discharge plasma*. The number of charges in the plasma over the number of the rest of neutral particles available at the given gas pressure is called the *degree of ionization*, and in the most processing plasmas, it is much less than one, often 0.01. Figure 2.1 schematically shows the above described processes of the phase transfers caused by increasing the delivered thermal energy (heating power) to the plasma state.

We can recognize several kinds of energy, i.e. mechanical energy ($mv^2/2$), electric energy (electron volt or eV), thermal energy (kT), or optical or photon energy ($h\nu$), where m is the mass, v is the velocity, e is the electron charge ($e = 1.6 \times 10^{-19}$ As), V is the electric potential (volts), k or $k_B$ is the Boltzmann constant ($k = 1.38 \times 10^{-23}$ W s/K), h is the Planck constant ($6.63 \times 10^{-34}$ kg m$^2$/s), and $\nu$ is the photon frequency. The necessary energy for the first ionization (i.e. for losing one electron) of a neutral particle is given by so-called the first ionization potential, whose value depends on the type of gas. The ionization potentials for different atoms and molecules are tabulated, and their typical values for the single (first) ionization are of the order of 1–10 eV. The lowest known potential has cesium (3.89 eV), the highest one has helium (24.59 eV). The values of all other elements are between Cs and He. Potentials for metals are lower than for gases, e.g. titanium (Ti) has 6.83 eV, and argon (Ar) has 15.76 eV.

If we compare the electric energy of 1 eV (i.e. with the potential of 1 V) to the thermal energy equivalent (kT), we will get a surprisingly high corresponding temperature:

$$T = eV/k = 1.6\times10^{-19}/1.38\times10^{-23}\left[\text{As V/VAs/K}\right] \approx 11600\ \text{K} \approx \mathbf{11300°C} \tag{2.1}$$

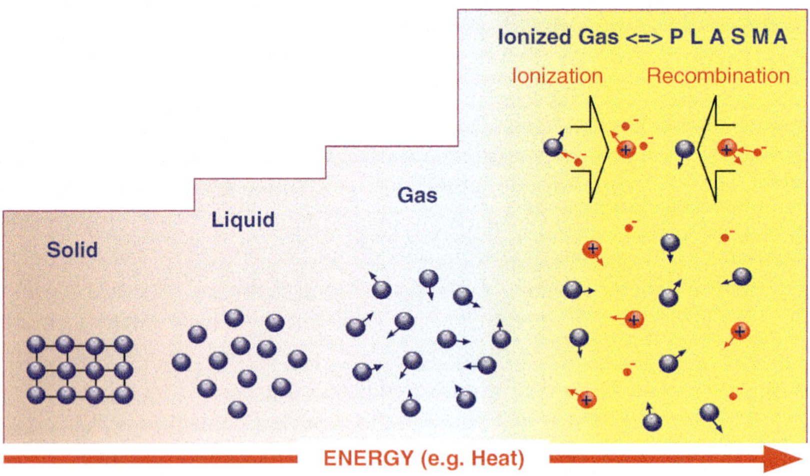


**Figure 2.1** Simple schematic description of the phases of matter and the formation of the plasma state.

This explains why our Sun, which is a gigantic plasma ball, has an extremely high temperature. The Sun's surface (its corona) is about 5600 °C, and at the center, the temperature is estimated to be 15 000 000 °C. In some practical applications, by using so-called thermal or hot plasmas generated by the high-power generators (about 1–10 MW), we can generate plasma temperatures of about 10 000 °C. Such arc type plasmas can be used in pyrolytic processes in metallurgy, waste incineration, etc. These temperatures exceed melting points of nearly all materials, and such hot plasmas are evidently not useful for processing of solid surfaces, particularly on the heat sensitive materials.

The visible light from the gas discharges is often considered a plasma. However, only parts of the gas discharges can be correctly considered plasma, e.g. the "positive column", which is a substantial part of the direct current (DC) gas discharge generated between an anode and cathode. Though this part can be considered the plasma, the more intense radiating parts visible in the discharges around the electrodes or on the reactor walls are not the plasma. The plasma can be considered only in those discharge regions where the density of the positive ions ($n_i$) is balanced by the negative electrons ($n_e$) if no negative ions are present. The plasma always keeps such electrical neutrality (the charge balance), and this important property is called the *quasineutrality* of the plasma:

$$n_e \approx n_i \tag{2.2}$$

The electrodes and the reactor walls represent disturbances in the plasma, and they intrude the electrical neutrality. However, the plasma always responds to such intruding regions by forming some kind of sheaths with an opposite charge to that of the disturbing surfaces. These *space-charge sheaths* are adjacent to the electrodes, where the electrically neutral plasma is insulated and separated from the charged surfaces. Any disturbance of the charge neutrality in the plasma is immediately screened by the opposite charges. Due to a substantial difference (about 1000 x) between the mass of electrons and ions, the light mobile electrons are always faster than ions on all surfaces of the passive objects immersed in the plasma and they form there a negative surface charge. The plasma responds by the so-called sheath above this negative surface with an equivalent positive charge at the plasma side. The mass difference between charges is important at the high frequency of microwaves, where ions are too heavy even to follow oscillations of the microwave electromagnetic field.

Keeping the charge balance around any object immersed in the plasma by screening this object by a "shell" of positive ions has a characteristic distance between the negative surface and the positive wall of the sheath, which is called Debye length ($\lambda_D$). This length (in meters) depends on the electron (or ion) density (n) (in $m^{-3}$) and on the electron energy $kT_e$, where the electron temperature $T_e$ is given in Kelvins:

$$\lambda_D = e^{-1}(\varepsilon_0 k T_e/n)^{1/2}[m] \Leftrightarrow \left[(A\,s)^{-1}\left(A\,s\,V^{-1}m^{-1}K\,VA\,s\,K^{-1}m^{-3}\right)^{1/2}\right] \quad (2.3)$$

Note the equivalence of units at both sides of the equation above: $\varepsilon_0$ is the permittivity of the vacuum or the free-space [$\varepsilon_0 = 8.85 \times 10^{-12}$ (A s $V^{-1}m^{-1}$)]. After introducing the constant values (e, $\varepsilon_0$ and k) a simplified form of the expression, see Eq. (2.3), is the following:

$$\lambda_D \approx 69\left(T_e/n\right)^{1/2}[m] \quad (2.4)$$

For example, the electrons with an electron voltage of 5 eV ($5.8 \times 10^4$ K) in a plasma with the density of $10^{16}$ $m^{-3}$ ($10^{10}$ $cm^{-3}$) will exhibit the Debye length of about $\lambda_D \approx 0.2$ mm. Due to this Debye screening effect, any plasma discharge can only be generated in spaces with the linear dimensions longer than the Debye length. In shorter distances, the plasma cannot exist.

As mentioned, the space-charge sheaths are usually visible as intense discharge parts surrounding the electrodes and on the reactor walls. The sheaths are usually thicker than several $\lambda_D$, (about 10–100 times), because the electrode surfaces exchange the charges with the plasma, and the thicknesses of created space-charge sheaths depend on the plasma parameters. For example, the sheath thickness depends roughly on $p^{-1/2}$ (where p is the gas pressure). The properties of the space-charge sheath also depend on the type of the power used for the plasma generation (DC or AC), on the power frequency, meaning radio frequency (RF), microwaves, etc., and on the level of the delivered power, which controls the plasma density. The sheath regions on the surfaces are often the most important parts during processing interactions with the plasmas because the final processes impact the resulting surface modifications.

## 2.2 Generation of the Plasma, Townsend Coefficients, Paschen Curve

The gas discharge plasma has been introduced without a description of the generation processes. The plasma generated by an increasing heat, shown in Figure 2.1, was a schematic simplification. A simple example of a slightly enhanced gas ionization is an ordinary gas flame, but the flame is still not generally considered a typical plasma. In Chapter 5, we will show that it is incorrect. The normal dry air is electrically insulating, and its breakdown requires an electric field of about 30 kV/cm.

In many practical applications, the electric or electromagnetic fields are applied to a gas by means of electrodes or by electromagnetic waves to generate sustaining discharge plasmas. This can be done, e.g. by emission of electrons from an electrically biased cathode in a DC circuit and accelerating these electrons by electric fields toward an anode to the ionization energies of the given gas or by energizing primary electrons present in the gas from other electron sources, e.g. from hot filaments. However, some electrons are always present around us due to ionization by the continuous cosmic radiation.

The basic experimental tests of generation of the steady-state gas discharge in the DC electric discharges were carried out by J.S. Townsend, see, e.g. Refs. [2.1, 2.2]. More fundamental knowledge and details were published after Townsend's work in several books on gas discharges, see, e.g. Refs. [2.3, 2.4]. Figure 2.2 shows an idealized scheme of the avalanche processes present during ionization of an atomic gas between electrodes. A similarly simple scheme can be used for molecular gases.

It is necessary first to explain the processes shown schematically in Figure 2.2. In Townsend's experiments, he introduced two main coefficients for the ionization probability: alpha ($\alpha$), which is the *1st Townsend coefficient* for ionization in a volume between DC electrodes separated by distance "d", and gamma ($\gamma$), which is the *3rd Townsend coefficient* for probability of emission of secondary electrons from the cathode during bombardment of the cathode surface by ions created by the volume ionization. The coefficient $\gamma$ is related to so-called *work function* of the given cathode material.

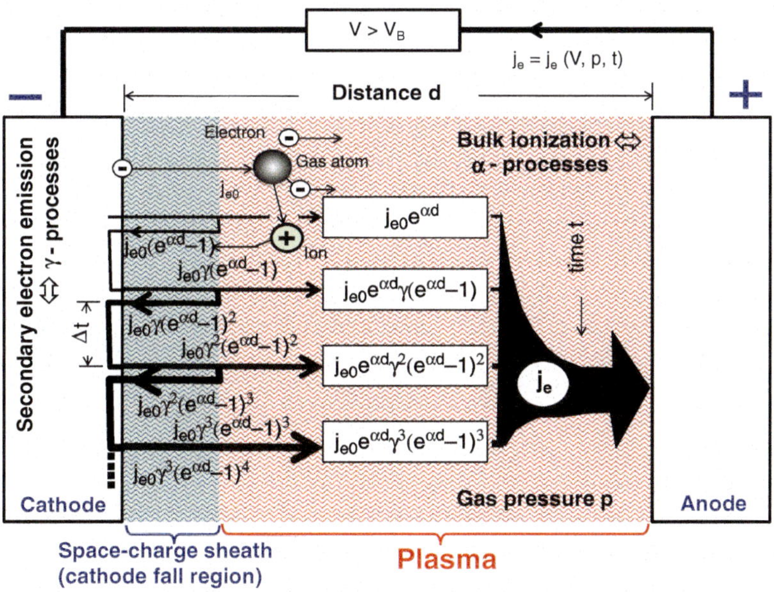


**Figure 2.2** Idealized scheme of an avalanche process of the gas breakdown in a DC system.

The Townsend coefficient $\alpha$ depends on the collisions and, therefore, on the gas pressure (p) between electrodes in the gap (d). Coefficient $\gamma$, also called the *secondary electron emission coefficient*, represents the generation of the secondary electrons by surface processes on the cathode. If the voltage between electrodes is high enough to ionize the gas, an initial flux of electrons ($j_{e0}$) emitted from the cathode is capable to ionize gas atoms (or molecules in a molecular gas) and create positive ions with a probability of $e^{\alpha d}$. Assuming that each new ion is formed by an existing electron emitted from the cathode, the resulting flux of positive ions will be $j_{e0}$ ($e^{\alpha d}-1$). Being positive, these ions will be forced back to the cathode, bombard the cathode surface, and generate additional (secondary) electrons with probability $\gamma$. The electron flux ready for the further ionization will then be $j_{e0}\ \gamma\ (e^{\alpha d}-1)$. Repetition of such processes can lead to an avalanche and result in an electron flux $j_e$, which is the current flowing in the DC circuit as in Figure 2.2.

If we consider only one initial electron in the process described above, instead of generation of $j_{e0}$, the simple expression will have following form:

$$1 = \gamma(e^{\alpha d} - 1). \tag{2.5}$$

Here, minus one means the initial electron, which was not balanced by ion. This expression represents the sustained ionization and forming a breakdown current in the plasma. The voltage necessary for the breakdown is called breakdown voltage ($V_B$).

Five important notices should be added to the simplified description of the breakdown showed in Figure 2.2 and expressed in equation Eq. (2.5):

1. Due to collisions, $V_B$ is usually higher than the first ionization potential of the gas used (roughly about 10–20 times).
2. The breakdown depends on the distance (d) and on the time (t) until the avalanche reaches a steady state. The value of $\Delta t$ in Figure 2.2 is the time interval necessary for the cycle: electron impact – ionization – ion hitting the cathode – secondary electron emission. Therefore, a more realistic approach includes time t: $\alpha = \alpha$ (d, p, t).
3. $V_B$ can be expressed as a function of $\alpha$ and $\gamma$, but because the Townsend coefficient $\alpha$ can be expressed in terms of gas pressure p, the resulting function can be $V_B = V_B$ (p.d, $\alpha$, $\gamma$), which forms so-called the Paschen curve.
4. The gas discharges are often classified according to the prevailing process of electron production as $\alpha$-discharges or $\gamma$-discharges.
5. J.S. Townsend considered the probability of the ion-based ionizations (2nd probability coefficient $\beta$), but such ionizations were found practically negligible.

Pioneering experiments carried by Friedrich Paschen (see Ref. [2.5]) revealed that $V_B$ depends on the distance (d) between electrodes and the gas pressure

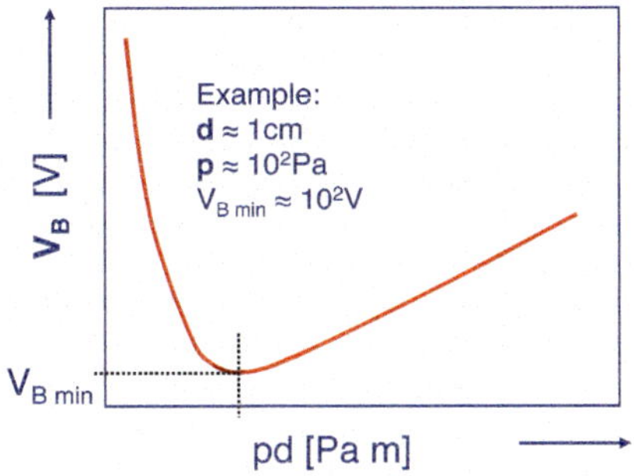


**Figure 2.3** Shape of the Paschen curve showing a minimum breakdown voltage for the tested gas. The example shown in the figure is for argon at 100 Pa (0.75 Torr) between 1 cm separation of the electrodes; the necessary breakdown voltage is about 100 V.

(p). The function $V_B = f(p\ d)$ is called *the Paschen curve*, and it always appears with a valley-type shape, as shown in Figure 2.3, similar for all gases. For constant distance (d), the shape of the Paschen curve can be explained by effects of the gas pressure. At low gas pressures, due to the low frequency of collisions in a corresponding low number of gas particles, the electrons need higher energy to ionize gas particles enough and sustain the discharge. Therefore, to generate and to sustain the low-pressure plasma, high breakdown voltages are often needed. At low pressures, the ionization process often requires some kind of additional pre-ionization, e.g. a hot filament for emission of additional electrons. On the opposite side, at high pressures, there is high collision frequency and, consequently, high recombination rates. These conditions again require higher voltages to start and sustain the discharge. Between the low and high pressures, a certain pressure interval always exists, where the $V_B$ is lowest in the voltage "valley." As mentioned, the $V_B$ of the dry air at atmospheric pressure in a gap of 1 cm is about 30 kV. At reduced pressures, it can be about 100 V = 0.1 kV, as shown in Figure 2.3.

## 2.3 Generation of the Plasma by AC Power, Plasma Frequency, Cut-off Density

It might be expected that the gas breakdown in AC fields oscillating with certain frequency is simply an oscillation of the DC electric field. However, such an approach is misleading because the oscillating field is not pure electric but is always an electromagnetic field. For example, the oscillating electromagnetic field can ionize the gas through the dielectric walls of a reactor without placing electrodes in direct contact with the gas discharge. This is not possible in the DC case. The difference between the mass of electron ($9.1 \times 10^{-31}$ kg) and ion (about $1.7 \times 10^{-27}$ kg for protons) causes a much higher mobility of electrons than of ions. At high AC frequencies in the microwave field, the motions of electrons and ions are, therefore, almost entirely decoupled. In the processing applications, the plasma generation by high frequency power, e.g. by microwave, is used more frequently than the simple DC power, because there is usually better plasma stability and higher plasma density. Moreover, an auxiliary DC bias can be used to control the motions and energy of the charges interacting with the processed substrates. A disadvantage is usually a lower power efficiency with a higher price of the AC generators than the DC generators.

Basic parameters of the electromagnetic wave are its frequency ($\omega = 2\pi$ f [Hz $= s^{-1}$]) and the power (only voltage and current amplitudes if we ignore the phase). The frequency can be expressed as an angular frequency $\omega$ (suitable in theoretical descriptions) or as a normal frequency f $= \omega/2\pi$. Different frequencies of incident waves lead to different generation effects. In a first approximation, disregarding the frequency effects, the AC generation of the plasma means an interaction of the electromagnetic waves of a given frequency with a weakly ionized (pre-ionized) gas. As mentioned, a weak pre-ionization of the gas is always present due to the cosmic rays. The power carried by the electromagnetic wave may be absorbed by the particle motions, e.g. by electrons. Moreover, an interaction of the incident electromagnetic wave with the plasma may often lead to the generation of other kinds of waves, which subsequently interact with the plasma particles and loose the energy. Some heavy particles can be ionized by such waves directly and by electrons produced in ionizations. Accelerated electrons (lighter and more mobile than ions) are then capable to ionize the neutral particles further and increase the plasma density and/or energy of other particles up to certain limit dependent on the type of gas and incident waves. The process of the energy transfer from the wave to the particles is often denoted as a *heating of the plasma* (particularly in plasma fusion experiments and in high-energy plasmas). The most rigorous way to describe the interactions of waves with plasma would be solving the system of differential equations consisting of four Maxwell equations for the electromagnetic waves and the Boltzmann differential equation for the movements of particles in the plasma, see, e.g. Refs. [2.4, 2.6]. Such an approach exceeds the aim and level of this book.

There are different mechanisms of the absorption of electromagnetic waves in the plasma. The two basic mechanisms are collision-based absorption and collisionless absorption. In the collision-dominated mechanism, the wave energy is transferred directly into the motions of plasma particles. Due to collisions, the energy acquired by these particles is continuously dissipated in the plasma volume, and the wave successively loses its energy along its path. A simplest example of the collision mechanism is the ohmic heating by DC power. A similar approach can be expected for alternating fields at low frequencies (f $<$ 10 kHz). An important parameter is naturally the power transported by the wave into the plasma. However, it is impossible to define high or low power without introducing the volume of its incidence or without knowledge on the number of particles on which the wave power is applied. In a rough estimation for the low-pressure plasmas (< $10^3$ Pa or 10 Torr), the power density exceeding 1 W/cm$^{-3}$ might be considered high. At moderate and low powers, the absorption of the wave energy and the density of the generated plasma could linearly follow the power level. On the contrary, at high powers, certain nonlinear effects can take place, which can result in substantially higher plasma densities than expected from the linear approach. However, the power dissipation in the plasma might be non-uniform, accompanied by a non-uniform distribution of the plasma density.

The value of an ultimate importance in the wave absorption mechanisms is the *plasma frequency*. As explained, the gas discharge plasma is a complex dynamic system of charges and activated neutrals. Due to the Coulomb forces between charges, the plasma exhibits specific oscillations with characteristic frequency called plasma frequency. The Coulomb force between charges $q_1$ and $q_2$ is expressed as follows:

$$F_{Coulomb} = k_C(q_1\, q_2)\, /\, r^2\, [Ws/m] \tag{2.6}$$

where $k_C$ is the Coulomb constant ($k_C = 9\ 10^9$ V m/A s) and "r" is the distance between these charges. The force between charges can be negative (attractive), e.g. electrons and ions, or can be positive (repulsive) for the same polarity of charges, e.g. pairs of electrons or ions. In an idealized approximation shown in Figure 2.4, the stationary systems of both types of charges will oscillate when

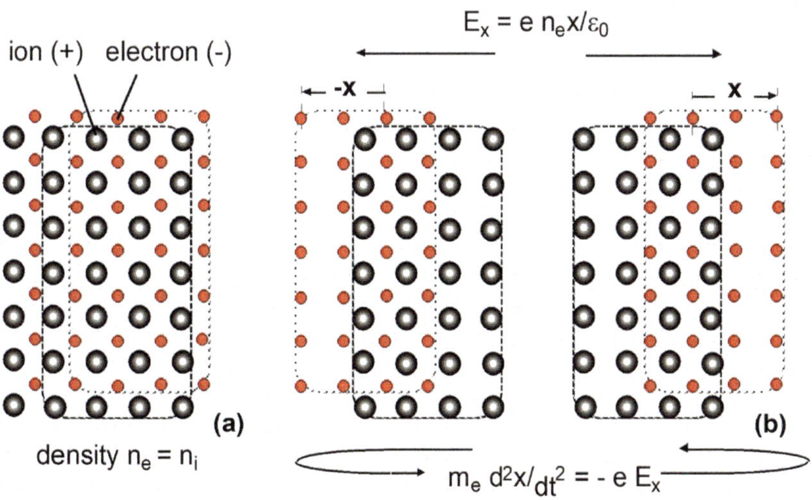


**Figure 2.4** Idealized description of the plasma oscillations. (a) The plasma as an electrically neutral system of electrons and ions. (b) Displacement of electrons from their positions to a small distance, designated x, forms an electric field $E_x$ and electrons start to oscillate.

pulled from an initial position to distance "x" and then released. The electric field formed by such displacement of electrons depends on the electron density ($n_e$) and the length (x) of their displacement, $E_x = e\ n_e\ x/\varepsilon_0$, where e is the charge of electron and $\varepsilon_0$ is the permittivity of vacuum. The oscillation frequency ($\omega_p$), derived from the equation for the harmonic oscillator assuming driving force of $eE_x$ (shown in Figure 2.4), is the plasma frequency. This frequency for electrons (*plasma electron frequency*) is the following:

$$\omega_{pe} = 2\pi\sqrt{\frac{ne^2}{\varepsilon_0 m_e}} \cong 18\pi\sqrt{n} \tag{2.7}$$

where $m_e$ is the electron mass ($9.11 \times 10^{-31}$ kg). The plasma frequency for ions with mass $m_i$ will be $(m_i/m_e)^{1/2}$ times smaller (i.e. about 100x).

In Eq. (2.7), the plasma frequency depends only on the square root of the electron (or ion) density n (the plasma density). The value calculated in an ordinary frequency f instead of its angular form $\omega$ will be $2\pi$ times smaller $f_{pe} = \omega_{pe}/2\pi$. For example, in a plasma with $n = 10^{16}\ m^{-3}$ the plasma frequency will be $f_{pe} \approx 900$ MHz.

For incident electromagnetic waves, the plasma looks like a complex disperse dielectric. "Disperse" means frequency dependent properties. The properties of plasma depend on the frequency of the incident wave. In plasma theory, the current density vector **j** in plasma depends on the plasma conductivity ($\sigma_p$). One can write $\mathbf{j} = \sigma_p\ \mathbf{E}$, which represents the Ohm's law. *The conductivity of plasma* is a complex value depending on the frequency $\omega$ of the incident wave and on the plasma electron frequency $\omega_{pe}$:

$$\sigma_p = \frac{\varepsilon_0 \omega_{pe}^2}{j\omega + v_{coll}} \quad \text{and} \quad \sigma_{dc} = \frac{e^2 n}{m_e v_{coll}} \quad \text{for the DC case}\left(\text{where } \omega = 0\right) \tag{2.8}$$

In this equation, j is an imaginary unit ($j^2 = -1$), $\varepsilon_0$ is the permittivity of the free space (vacuum), the plasma frequency ($\omega_{pe}$), introduced in Eq. (2.7), $m_e$ is the electron mass, and $\nu_{coll}$ is the collision frequency. It is seen that the conductivity ($\sigma_p$) of a plasma depends linearly on the plasma density (because $\omega_{pe}^2 \approx n_e$) and depends inversely on the collision frequency, which "slows down" the particles. The effect of the $\nu_{coll}$ is seen in the second part of Eq. (2.8) for the conductivity in the DC case, where the frequency of the power is $\omega = 0$.

Because *the plasma is a complex disperse dielectric*, it is possible, similar to other dielectrics, to introduce a dielectric permittivity ($\varepsilon_p$) of the plasma:

$$\varepsilon_p = \varepsilon_0\left[1 - \frac{\omega_{pe}^2}{\omega\left(\omega - j v_{coll}\right)}\right] \approx \varepsilon_0\left(1 - \frac{\omega_{pe}^2}{\omega^2}\right) \tag{2.9}$$

An approximation on the right side of Eq. (2.9) is for the low pressures, where the $\nu_{coll}$ can be neglected in comparison with the wave frequency: $\omega \gg \nu_{coll}$.

In Eq. (2.9), two general possibilities exist:

1. The wave frequency $\omega$ is greater than the plasma frequency ($\omega > \omega_{pe}$). There, the dielectric permittivity is positive, $\varepsilon_p > 0$. In this case, the wave can penetrate into the plasma, lose its energy, and "heat" the plasma.
2. The wave frequency is lower than or equal to the plasma frequency ($\omega \leq \omega_{pe}$); then the dielectric permittivity of the plasma would be negative, $\varepsilon_p \leq 0$, which has no physical sense. In this case, the only possibility is the wave cannot penetrate into the plasma and it is reflected.

Consequently, until the plasma does not reach certain critical value ($n_{crit}$) of its density, the wave can penetrate into such plasma and deliver its power. The value of $n_{crit}$, which is also called the *cut-off density*, can be derived from Eqs. (2.7 and 2.9):

$$n_{crit} = 1.24 \text{ x}10^{-2} f^2 \tag{2.10}$$

where f is the wave frequency ($f = \omega/2\pi$). Whenever the plasma density (n) approaches $n_{crit}$, the plasma starts to reflect the wave. These cases can be expressed through the following frequencies:

$$\omega^2 = \omega_{crit}^2 \approx n_{crit} > \omega_{pe}^2 \approx n(\text{transition}), \text{or}\, \omega^2 = \omega_{crit}^2 \approx n_{crit} \leq \omega_{pe}^2 \approx n(\text{reflection}). \tag{2.11}$$

Eq. (2.11) implies an important conclusion: *the higher the frequency of the wave, the higher the density of plasma can be generated by the power delivered by this wave.* Therefore, the high-frequency systems are more suitable for generation of the high-density plasmas then low-frequency ones. The critical density for the microwave power ($f_{MW} = 2.4$ GHz) should be $n_{crit} \approx 7 \text{ x } 10^{10}$ $cm^{-3}$; for typical radio frequency power ($f_{RF} = 13.56$ MHz), it is only $n_{crit} \approx 2 \text{ x } 10^6$ $cm^{-3}$. Optical lasers can generate high-density plasmas, and the gamma rays can generate the highest density in the nuclear processes.

The plasma density required for the processing applications is usually higher than the $n_{crit}$ in certain regions of the plasma. This is possible because the theoretical approach explained above is based only on a linear approximation, assuming low or moderate power transported by the wave under small collisions ($\omega \gg \nu_{coll}$). Increasing the power substantially is possible to push the wave into the plasma despite reflections. In this case, the linear approach does not work, and the plasma density may exceed critical value but at the cost of considerable losses in the generation efficiency. Moreover, the plasma density along the path of the wave is non-uniform with higher density at the incidence area with a steep decrease towards the bulk of the plasma. Therefore, other methods must be applied in generation of uniform overcritical (overdense) plasmas in large volumes, or over large surface areas. For example, suitable methods are based on the wave resonances as explained below.

The plasma may exhibit identical conditions for the incident waves in all directions, like in isotropic dielectrics, but it may also have certain preferred directions, caused by auxiliary external forces similar to the anisotropic dielectrics. Such anisotropic plasmas are used in resonant processing systems in special nonlinear regimes of the wave absorption or in the generation of highly overcritical plasma densities. Two most typical regimes are the *Landau damping* and the *electron cyclotron resonance* (ECR). The plasma anisotropy in both these regimes is created by a static magnetic field (induction vector **B**) applied in the plasma region. The plasma properties differ along this field compared to other directions. In both cases, the wave power is absorbed through mechanisms independent on the particle collisions, and they are denoted as the *collisionless absorption*. The collisionless plasma can be generated at low gas pressures ($\leq 1$ Pa $\approx 8$ mTorr), where long mean-free paths (MFPs) allow better control of the directions and energies of the charged particles.

Let us assume the planar electromagnetic right-hand polarized (RHP) wave having the intensity of electric field [E(t)] and the intensity of magnetic field [H(t)] [or the induction B(t) = $\mu$ H(t), where $\mu$ is permeability] oscillating in the planes perpendicular to the wave velocity, as shown in Figure 2.5. The left-hand polarized (LHP) waves in Figure 2.5 would have opposite directions of the angular motions. The high-frequency wave moving in the plasma along the stationary magnetic field with the induction vector **B** interacts mainly with the plasma electrons; ions are too slow for the interactions. Because of the electrons' high mobility, they can be trapped and carried by the wave in the

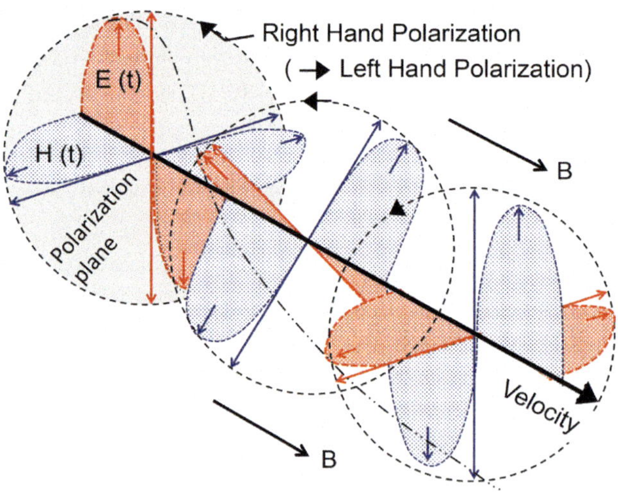


**Figure 2.5** Schematic representation of a planar RHP electromagnetic wave moving in the direction of static magnetic field of induction B. Intensity of the magnetic field is H = B/μ, where μ is the permeability.

electrostatic "valleys" of the wave. The electrons are accelerated by the wave if their phase velocity is in suitable relation with the phase velocity of the wave. In this Landau damping regime, the wave transfers its energy to the electrons, meaning it is dumped.

In a simple approach, this process of electron acceleration in Landau damping may be compared to a surfer on the sea waves who must have the right starting velocity to be properly carried and accelerated by the sea waves (according to F.F. Chen in Ref. [2.6]). The high-energy electrons generated by the Landau damping enhance the overall ionization in the plasma, independent on the critical plasma density. This is considered to be the main mechanism responsible for the high-density RF plasma generation in helicon plasma sources, see Ref. [2.7]. Experiments with the helicon plasma sources confirm that the absorption mechanisms are not very sensitive to the value of induction **B** of the static magnetic field, see Refs. [2.8–2.10]. Typical values are B ≈ 20–30 mTesla, but the whole interval of inductions between 0.1–500 mTesla have been examined successfully.

An important collisionless absorption of the microwave power is the ECR, where electrons having mass $m_e$ and the velocity component $v_{e(norm)}$ normal to the vector **B** of the static magnetic field (induction $B = |\mathbf{B}|$) are forced to gyrate around the magnetic flux lines at Larmor radius ($r_L$) defined as the following:

$$r_L = m_e v_{e(norm)} \left(e\,B\right)^{-1} \tag{2.12}$$

The cyclotron or gyration frequency for electrons is expressed as the following:

$$\omega_{ce} = e\,B / m_e \tag{2.13}$$

This frequency can be expressed for ions by replacing the mass $m_e$ by the ion mass $m_i$. The condition for the incident wave with angular frequency ω (= 2πf) to undergo resonant absorption of its energy by the electron resonances, called ECR, is the following:

$$\omega = \omega_{ce} \text{ or } f = e\,B / (2\pi m_e) \tag{2.14}$$

The Landau damping and the ECR have four similarities:

1. Both are based on electron interactions.
2. Both are based on the waves moving along the static magnetic field.
3. Both need auxiliary magnetic field.
4. Both are independent on collisions.

However, on the contrary to the Landau damping the ECR mechanism is sensitive on the magnetic induction B. The ECR regime needs magnetic field of induction $B = B_{ce}$, where:

$$B_{ce} = \omega m_e / e \tag{2.15}$$

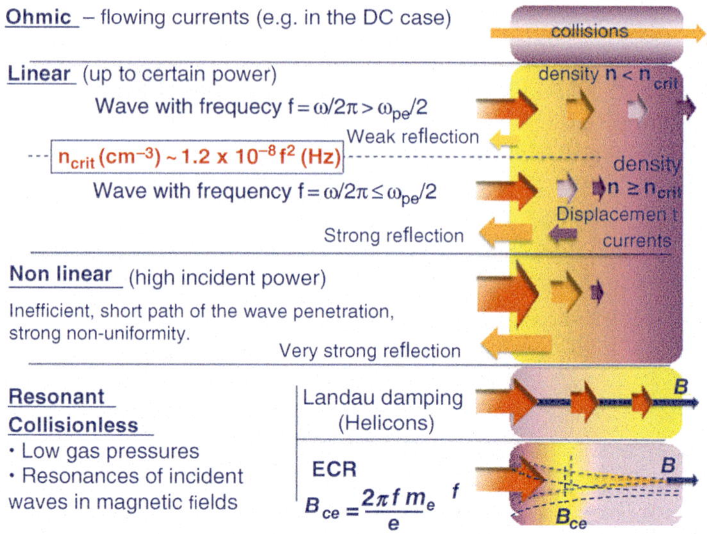


**Figure 2.6** Graphical illustration of the plasma generation by incident electromagnetic waves.

Typical value $B_{ce}$ for the microwave generation (f = 2.4 GHz) is $B_{ce}$ = 8.57 $10^{-2}$ T = 857 Gauss. For 2.35 GHz the resonance requires magnetic field of B = 8.4 x $10^{-2}$ T = 840 Gauss. This level of the magnetic induction can be generated by the electromagnetic coils or by permanent magnets. Even though the required magnetic induction for lower frequencies are lower (for the RF f = 13.56 MHz it is only $B_{ce}$ = 4.8 Gauss), the most commercial plasma systems using ECR regimes are powered by the microwaves where the critical frequency is relatively high. The energy transported by the wave in the ECR is absorbed in the plasma efficiently and independently on the collisions. The high-density ECR microwave plasma is easy to ignite and it has considerable stability. Schematic Figure 2.6 illustrates all AC generation processes of the plasma described above.

## 2.4 Space-charge Sheaths at Different Frequencies of the Incident Power

The schematic illustration of the characteristic regions in the DC discharge between electrodes is shown in Figure 2.7.

The applied voltage ($V_{DC}$) causes motions of the charged particles in the generated DC current. As explained in Section 2.1, the equivalent number of electrons and ions (i.e. quasineutrality) can be considered only in the positive

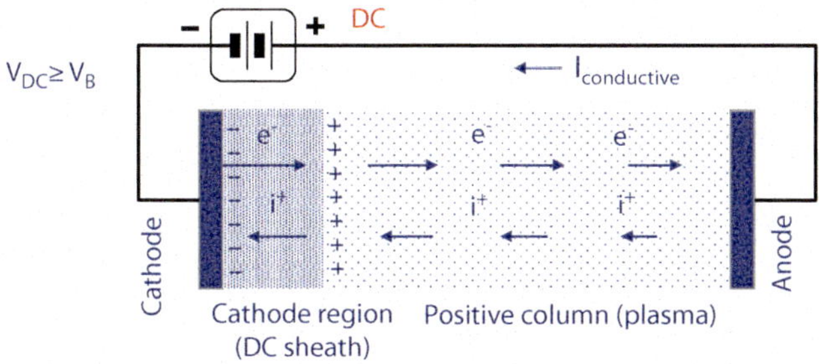


**Figure 2.7** Schematic representation of the electron and ion currents and important discharge regions in the DC plasma. The cathode fall region represents the space-charge sheath.

column, which is considered as the plasma. Because the current density ($j_{DC}$) must be the same everywhere in the DC circuit, the current can be expressed as a sum of the electron and the ion current densities:

$$j_{DC} = \sigma_p E = n_e q_e v_e + n_i q_i v_i = n\,e\left(v_e + v_i\right) \tag{2.16}$$

The value $\sigma_p$ represents the conductivity of the DC plasma, see Eq. (2.8). The electron and ion charges are identical $|q_e| = |q_i| = e$, but their charges have opposite signs. In the electric field, the velocity of ions $v_i$ has an opposite direction, but the ion charge is opposite to the electron. The current density depends on the density of transported charged particles, their charge (single charged ions), and the velocities (energies) of charged particles. Assuming $v_e \gg v_i$, Eq. (2.16) can be simplified to $j_{DC} \approx n\,e\,v_e$.

As shown in Figure 2.7, the electrons emitted from the cathode are accelerated toward the anode through the cathode dark space or cathode fall (the DC form of the space-charge sheath) and they should acquire enough velocity (energy) to ionize the gas. Therefore, the cathode fall is followed by the area of the first ionization, which is accompanied by an intense optical emission. Due to ionizing collisions, the electrons lose part of their energy (in Faraday dark space, which is not shown), and they continue in a steady flow along the positive column (plasma) to the anode, being compensated by an opposite flow of ions to the cathode. In the region close to the cathode, the ions are unable to follow fast electrons, and they screen the negative cathode surface. Therefore, cathode region is permanently electron deficient, regardless of a diffuse flow of ions. The main characteristics of the DC circuit are the following:

- Electrodes are in contact with the discharge.
- The gas breakdown depends on the cathode emission of electrons.
- Electrode phenomena (secondary emission, sputtering, melting) are important.
- Conductive (ohmic) currents are present.

- Stationary cathode fall-region and its thickness (pressure, voltage, gas dependent) is another characteristic.
- Plasma density relates to the DC current.

Figure 2.8 shows a simplified schematic illustration and characteristic regions in an AC discharge between electrodes, which can be placed outside a dielectric chamber.

When an AC voltage is applied between two electrodes, the difference in the mobility of electrons and ions becomes important. At low frequencies, both electrodes in the AC circuit are cathodes during part of the AC cycle, and therefore, both electrodes are temporarily attached to the electron-deficient regions based on similar principles as in the DC case. In an opposite polarity, the negative charge on an electrode is compensated until the start of the next AC polarity, etc. The regions surrounding the electrodes are known as the space-charge sheaths. The oscillating field can generate space-charges inside the volume of gas. These charges can simplify the AC breakdown. The amplitude of the AC voltage necessary for generating and sustaining the discharge is usually higher at low frequencies and lower at higher frequencies, see Refs. [2.11, 2.12]. The simplified Figure 2.8 below represents a low-frequency AC case in the kHz range. The higher frequencies, in the range of RF and microwaves, cause different effects, as explained below.

The main characteristics of an oscillating electromagnetic field in an AC circuit are:

- The possibility of the electrodeless generation of the plasma exists (no ohmic contact of the generating electrodes with the plasma).
- The gas breakdown can start inside the volume.
- Forming of the space-charges inside the gas volume enables lower breakdown voltages than in the DC case.
- The sheath thickness and voltage oscillate with the AC frequency ω.

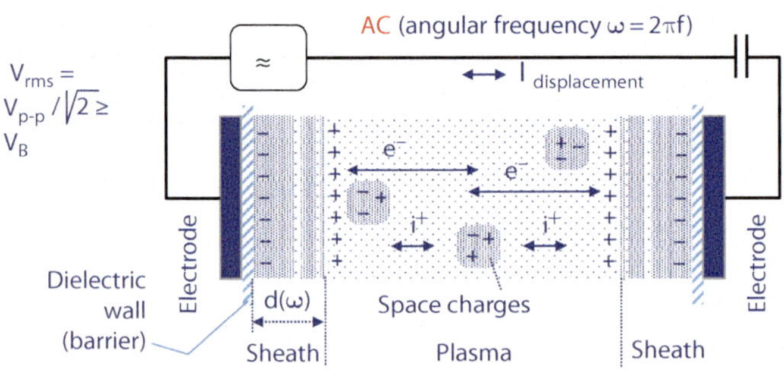


**Figure 2.8** Schematic representation of the electron and ion currents, space-charges, space-charge sheaths, and motions of charged particles in an AC plasma.

- At frequencies $\omega > \omega_{pi}$, the movements of electrons and ions are decoupled.
- The wave resonance phenomena allow generation of the high-density plasmas.

Properties of the space-charge sheaths depend on the gas, gas pressure, applied power, etc. Due to different masses of the electrons and ions, an important factor is the frequency of applied power. In the RF discharges with typical frequency order of $f = \omega/2\pi \approx 1–100$ MHz (see Figure 1.1, Chapter 1), the space-charge sheath can be described as a parallel-plate capacitor, where one plate is the plasma-sheath boundary and the second plate is the surface of the RF electrode. The RF sheath region between these plates is similar to a dielectric because it contains (theoretically) no charges. The electrons emitted from the powered or live RF electrode (always cathodic with respect to the plasma) quickly pass the space-charge region, but the ions are too slow to follow them and they form a positive boundary of the space-charge sheath on the side of the plasma to mirror the negative charge of the electrode.

The characteristic of the sheath region depends on the gas pressure. If the pressure in the plasma system is high (roughly > 1 Torr = 133 Pa), the electrons will collide with the neutral gas particles and generate ions inside the sheath region. These ions enhance secondary electron emissions from the electrode. The sheath character is changed from a weak $\alpha$-type (almost dielectric sheath) to a strong $\alpha$-type (volume ionizations) and $\gamma$-type (production of secondary electrons). These changes affect the whole discharge, see, e.g. Ref. [2.13].

Properties of the sheaths depend on the number of charged particles (electrons and ions) participating in the motions. Theoretical analysis shows that when the generator frequency $\omega$ is lower than the plasma frequency for ions $\omega_{pi}$ ($\omega_{pi} = \omega_{pe} (m_e/m_i)^{1/2}$, see Eq. (2.7), the motion of electrons and ions is coupled together, and they can react simultaneously. With respect to Figure 2.8, such a sheath in a low-frequency AC-powered discharge can be described in a similar way as with the DC discharges, with a periodically changing anode and cathode. Therefore, at $\omega \ll \omega_{pi}$, the displacement currents across the sheath can be neglected and, the sheath remains rather resistive. On the contrary, in the RF discharge, the motions of electrons and ions are fully decoupled and the corresponding space-charge sheath is denoted as an *ion sheath*. This happens for frequency $\omega$:

$$\omega_{pi} < \omega < \omega_{pe} \tag{2.17}$$

The ions are grouped at the plasma side at the sheath edge because they respond only to the time-averaged RF field and not to the oscillations of the RF field. The electrons inertia is neglected. The surfaces of small RF electrodes can acquire high negative potential with respect to the plasma. The time-averaged value of the potential $V_{dc}$ is called *self-bias*, see Ref. [2.14]. The voltages over the sheaths in the RF plasma oscillate with the frequency of the applied RF field. The time-dependent total voltage over the sheath $V_s(t)$ is the sum of the voltage $V_{dc}$ and oscillating amplitude of the RF voltage $V_{rf}$ applied to the sheath, Refs. [2.14–2.16]:

$$V_s(t) = V_{dc} + V_{rf} \cos \omega t \tag{2.18}$$

However, the voltage $V_{dc}$ also depends on the $V_{rf}$, see Ref. [2.14]:

$$V_{dc} = V_f - \frac{kT_e}{e} \ln I_0 \left( \frac{eV_{rf}}{kT_e} \right) \text{ where } V_f = \frac{kT_e}{2e} \ln \left( 4\pi\delta^2 \frac{m_e}{m_i} \right) \tag{2.19}$$

$I_0$ is the modified Bessel function of zero order, $m_e$ and $m_i$ are electron and ion masses respectively, δ is an eigenvalue, which represents the ion current to the wall (assume δ = 1). For small amplitudes of $V_{rf}$, the voltage $V_{dc}$ equals the *floating potential* ($V_{dc} \approx V_f$). The floating potential is the potential measured on an unbiased electrode floating in the plasma with respect to the ground (zero) potential. Often, the RF potential $V_{rf} \gg kT_e/e$ (where $T_e$ is the electron temperature, k is the Boltzmann constant, and e is the electron charge), and $V_{dc} \approx -V_{rf}$, see Ref. [2.14]. The sheath voltage is $|V_s(t)| \gg kT_e/e$ for most of the RF cycle. Because the average ion and electron currents to the wall must be balanced, during certain parts of the RF cycle, the sheath must collapse to allow electrons to escape. This happens when $|V_s(t)| \leq kT_e/e$ and a charge balance $n_e \approx n_i$ is settled throughout the collapsed sheath, see Ref. [2.16].

The *small electrode always exhibits higher self-bias than the larger electrodes at zero potential* (e.g. the grounded metallic reactor chamber). The potential drop between the plasma and the grounded electrode is the *plasma potential* ($V_p$). The distributions of the voltages $V_s$ and $V_p$ at the sheaths of electrodes with respective sizes $A_{rf}$ and $A_0$ were studied using different theories, see Refs. [2.17, 2.18]. The only approach supported by the experiments is based on the conservation of the positive ion current $j_{sh}^+ A_{rf} = j_0^+ A_0$, which gives the following important relation in the RF discharges:

$$V_s / V_p = (A_0 / A_{rf})^2 \tag{2.20}$$

This means that *the smaller the RF electrode is in capacitively coupled RF system, the higher the self-bias voltage is across its sheath*. This effect of the electrode size is important in designs and applications of capacitively coupled RF plasma systems and in cases when auxiliary RF biasing is applied, e.g. on the processed substrates in the microwave plasma.

Calculating the sheath thickness is possible from the Child-Langmuir theory, but an estimated sheath edge may misplace as much as 40% of the total space charge outside the sheath. One can derive the sheath thickness also by including almost an entire space-charge present in the sheath, see Ref. [2.19].

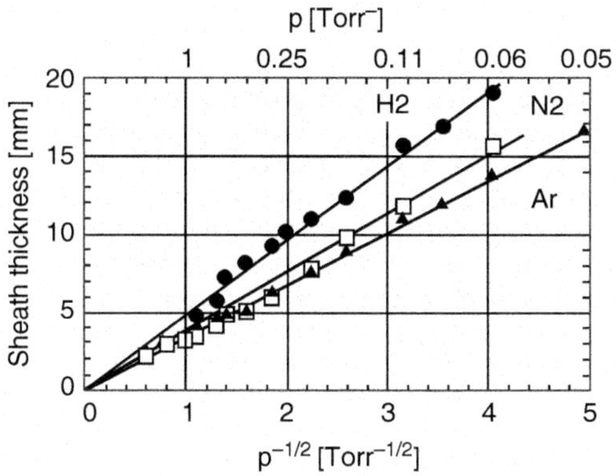


**Figure 2.9** The sheath thickness in the RF discharge in hydrogen, nitrogen, and argon as a function of pressure p. The linear dependence on $p^{-1/2}$ fits with the experimental data. (Data used from Figure 1 in Ref. [2.20]).

However, it might be best to utilize experimental factors affecting the sheath thickness. Experiments confirm that the thickness $d_s$ of the sheath in the RF discharges depends on the gas pressure p on the applied voltage $V_s$ and on the ion current $j_s^+$ density. According to Ref. [2.20], $d_s$ can be expressed as follows:

$$d_s \sim (1/p)^{1/5}, d_s \sim V_s^{3/5}, d_s \sim (1/j_s^+)^{2/5} \text{ for collisional } \alpha - \text{sheath} (d_s > l_i) \quad (2.21)$$

$$d_s \neq f(p),\ d_s \sim V_s^{3/4},\ d_s \sim \left(1/j_s^+\right)^{1/2} \quad \text{for collisionless sheath} \left(d_s < l_i\right) \quad (2.22)$$

where $l_i$ is the ion MFP. If the sheath is thinner than the MFP of ions, there are no collisions and the sheath is collisionless. Experiments in the RF plasma at different gas pressures, see Ref. [2.20], proved that the most reliable dependence is the following:

$$d_s \approx (1/p)^{1/2} \quad (2.23)$$

The experimental data for hydrogen ($H_2$), nitrogen ($N_2$), and argon (Ar) fit well with Eq. (2.23) as seen in Figure 2.9.

## 2.5 Classification of Gas Discharge Plasmas, Effects of Gas Pressure, Microwave Generation of Plasmas

### 2.5.1 Classification of Gas Discharge Plasmas

The gas discharge plasmas may be characterized in many categories or groups, either according to the parameters of their generation, or according to their characteristic properties. We can roughly recognize the following three classifications:

1. According to the generation parameters – high-pressure or low-pressure discharges, high-frequency plasmas (e.g. radio frequency – RF plasma, microwave plasma), DC plasma, pulsed DC plasma, magnetron sputtering plasmas, etc.
2. According to the characteristic properties – high-density plasmas, high or low particle energy plasmas, glow and arc discharges, etc.
3. According to the applicability – a suitable classification, which to some extent includes classifications (1) and (2), includes two basic groups:
   (I) equilibrium or "thermal" plasma
   (II) non-equilibrium or "cold" plasma

Classification (3) is based on different particle energies (temperatures) in the gas discharge plasmas. In the equilibrium (hot or thermal) plasmas, all particles have energies at the same level at thermal equilibrium due to particle interactions with energy exchanges. Most thermal plasmas are generated at the atmospheric or higher gas pressures where the high particle density [$\geq 10^{25}$ m$^{-3}$ ($10^{19}$ cm$^{-3}$)] is accompanied by the high collision frequency. Due to high temperatures ($\approx 10^4$ °C), e.g. in the high-power arc torches, such plasmas are not useful for treatments of the most material surfaces. However, at high gas pressures, generating non-equilibrium (cold or non-thermal) plasmas is possible. Principles and suitable systems will be described in Chapter 5.

Even though at special conditions (e.g. power, pressure), it is possible to ionize all gas particles and generate *fully ionized plasma*, in most cases, only a certain part of the gas particles present in the plasma is ionized. For these *partly ionized plasmas*, we can write the following:

$$n_e = n_i = n << n_{gas} \tag{2.24}$$

where $n_e$ is the density of electrons, $n_i$ is the density of ions, and $n_{gas}$ is the density of neutral gas particles at a given pressure and temperature. Defining the value of the *degree of ionization* ($\delta_i$) is useful and represents the ratio of density of charged particles to the density of all gas particles available at a given gas pressure. In terms of Eq. (2.24), the degree of ionization $\delta_i$ can be defined as the following:

$$\delta_i = n / n_{gas} \tag{2.25}$$

In the "ordinary" frequently used plasmas, the degree of ionization reaches values of $\delta_i \leq 0.01$ (i.e., only ≤1% ionization). In the microwave plasmas, this can be higher ($\delta_i \approx 0.1$). In the high-energy plasmas generated in fusion devices (e.g. Tokamaks, Stellarators, magnetic mirrors), working, e.g. with deuterium, the plasma can be fully ionized ($\delta_i \approx 1$). The differences between hot and cold plasmas are given mainly by the differences in the temperature of the prevailing amount of neutral (not ionized) gas particles in the plasma, i.e. $n_{gas} - n$, where $n = n_{e,i}$. An example of the lightning flash with an air temperature of ≈

30 000 °C belongs, of course, to the hot plasma category. The ionized gas in the discharge tubes and in light bulbs with walls almost at room temperature represents the cold plasma. The natural reason for such temperature differences is usually based on different collision frequencies of the gas particles $\nu_{coll}$ and the corresponding MFPs of these particles. These values depend on the gas pressure and on the gas temperature, gas composition, and the plasma generation parameters.

Typical densities of electrons (and ions) and the corresponding electron energies in different kinds of plasmas are shown in simple diagram in Figure 2.10. The energy of 1 eV corresponds to $T_e = 1.16\ 10^4$ K, (or $1.13\ 10^4$ °C). Figure 2.10 shows examples of the plasma sources used in different applications. Specific regions of the plasma parameters in the systems in this book, particularly the microwave plasma (denoted in Figure 2.10 as μW) and the ECR plasma, are emphasized in Figure 2.10 by the patterned areas with the full-line frames.

Even hot parts of flames have plasma characteristics. Recent measurements by double Langmuir probes in a premixed methane flame revealed as much as $10^{11}$ cm$^{-3}$ ions and an electron temperature around 0.2–0.3 eV, see Ref. [2.21].

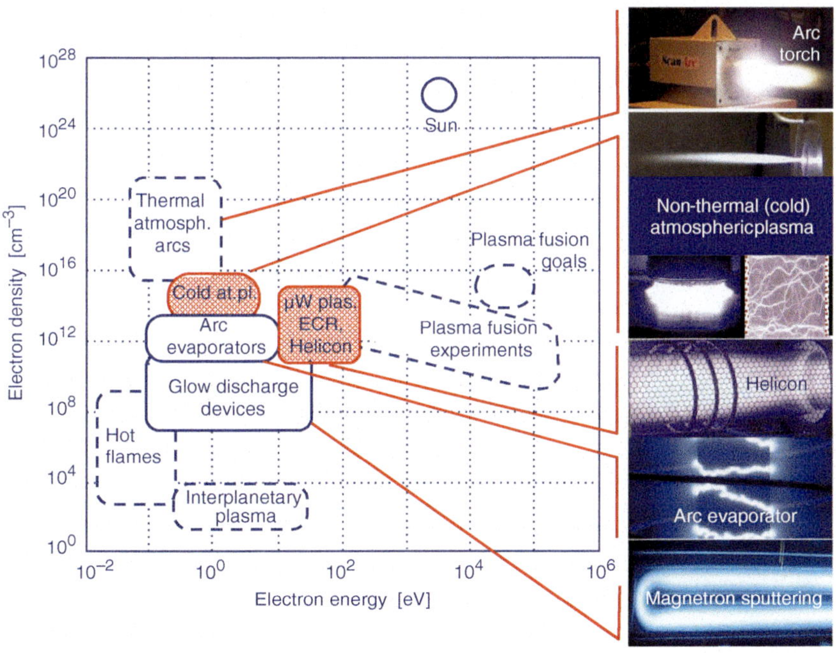


**Figure 2.10** Comparisons of typical electron densities and energies in plasmas. The non-thermal plasma systems described in this book are in patterned areas.

As the combustion is an atmospheric pressure process where the number of neutral molecules at a higher flame temperature can reach up about $10^{20}$ $cm^{-3}$ (see the next part), such plasma is only weakly ionized and prevailing parts of the neutral gas can be heated to several thousand °C. In Section 5.5, Chapter 5, we will show how the microwave plasma can affect the flames.

### 2.5.2 Effects of the Gas Pressure on Particle Collisions in the Plasma

Several effects of the gas pressure (e.g. collisions, Paschen curve, sheath thickness) have already been explained. The pressure and temperature dependencies of the density $n_{gas}$ of neutral gas molecules or atoms in the molecular or atomic gas can be estimated from Avogadro's law by following equation for several useful units of the gas pressure p:

$$\begin{aligned} n_{gas}\left[m^{-3}\right] &= 7.2 10^{22} p\left[Pa\right] / \left(T\left[°C\right] + 273\right) = 9.6 10^{24} p\left[Torr\right] / \left(T\left[°C\right] + 273\right) = \\ &= 7.2 10^{24} p\left[mBar\right] / \left(T\left[°C\right] + 273\right) \end{aligned} \tag{2.26}$$

The temperature T in the denominator of Eq. (2.26) is given in °C for T + 273, with the absolute temperature T in Kelvins. At the gas pressure of 1 mTorr and at room temperature (20 °C => 293 K), the gas density should be $n_{gas} = 9.6\ 10^{21}$ $/293 \approx 3.3\ 10^{19}\ m^{-3} = 3.3\ 10^{13}\ cm^{-3}$. The neutral gas particle density at 1mTorr is in the order of $10^{13}$ $cm^{-3}$.

A feasible way of generating a non-equilibrium plasma, i.e. with different particle energies, is to decrease the collision frequency through decreasing of the gas pressure. The collision frequency can be expressed as the following:

$$\nu_{coll}\left[s^{-1}\right] = \frac{\text{velocity of the considered particle } v_p[m s^{-1}]}{\text{mean-free path of the particle } l_p[m]} \tag{2.27}$$

For example, in ambient air at room temperature, where the density of neutral particles reaches the order of $10^{25}$ $m^{-3}$ ($10^{19}$ $cm^{-3}$), the collision frequency between molecules can be estimated to $v_{coll} \approx 10^9 - 10^{10}\ s^{-1}$ (see e.g. Figure 12 in Ref. [2.22]). At these conditions, the MFP $l_p$ between two collisions is roughly only $10^{-7}$ m (0.1 μm). Consequently, the interactions between particles are so frequent that any attempt by any particle to adopt more energy fails quickly (in $10^{-10}$ s = 0.1 ns = 100 ps).

In a first approximation, the velocity $v_p$ of particles in Eq, (2.27) depends only on external forces, e.g. on heat or on an applied voltage assuming the charged particles. Decreasing the particle velocity substantially is usually not simple, and increasing the velocities (energy) is often more desirable to decreasing them. However, an increase in the $v_p$ would cause an increase in collision frequency $\nu_{coll}$. Affecting the MFP of the particles becomes possible. In a simplest hard-sphere model the MFP $l_p$ of a particle with diameter $d_p$ is inversely proportional to the density of neutral gas $n_{gas}$, in which the particle is moving:

$$l_p\,(\text{in a gas})\,[m] \approx \left(Q_{p-g} n_{gas}\right)^{-1} = (\sqrt{2}\pi\, d_p{}^2 [m^2]\, n_{gas} [m^{-3}])^{-1} \tag{2.28}$$

$Q_{p\text{-}g}$ is the collision cross-section of the considered gas particle, which is characteristic for a given gas. Its value depends generally on the particle energy, but in a first approximation, it can be estimated by an *effective geometrical cross-section* ($2^{1/2}\ \pi\ d_p{}^2$) of the particle and consider it constant. For air molecules (diameter $d_p \approx 3.7 \times 10^{-10}$ m = 0.37 nm), $Q_{p\text{-}g} \approx 1.6\ 10^{-19}\ m^2$. For air at 1 mTorr pressure and 293 K (20 °C) the molecular density will count $n_{gas} \approx 3.3 \times 10^{19}\ m^{-3}$. Eq. (2.28) gives MFP of $l_p \approx 0.18$ m = 18 cm. Note that electrons should have approximately 4 times longer MFP, i.e. about 0.8 m at 1 mTorr.

From Eqs. (2.27 and 2.28), the collision frequency depends inversely on the particle density, and the particle density depends inversely on the temperature. Therefore, the MFP should grow with the temperature; at elevated temperatures of the gas, the free paths can easily reach multiple values. However, the calculations above are rather theoretical. For practical uses, one can rely on the multiple experimental observations to estimate the MFP $l_p$ for air at room temperature from the following expression:

$$l_p\left(\text{in air}\right)\left[mm\right] \approx 6.7 / \left(p\left[Pa\right]\right) \approx 50 / \left(p\left[mTorr\right]\right) \tag{2.29}$$

The MFP in air at 1 mTorr and room temperature is about 5 cm. At 350 °C, the MFP is about twice of the room value, i.e. appr. 10 cm.

Contrary to hot plasmas, the parameters of cold plasmas can be tuned for the treatment of surfaces of almost all materials and for treatments of biological objects like the human skin. Such plasma enables substrate temperatures as low as 50–70 °C during the treatment, e.g. during the growth of films where the classical chemical methods would require temperatures close to 1000°C. There are no temperature limits so far defined or commonly accepted in the literature for the characterization of cold plasma. Therefore, we can define this limit with respect to the material processing:

*The partly ionized gas discharge plasma with the neutral gas temperature up to 1000 °C may be considered a cold plasma, the temperatures over 1000 °C may be considered a hot plasma.*

Chemical reactions in cold plasmas are based on the collisions of electrons with gas molecules or atoms, which can become chemically reactive, e.g. ions, radicals, excited atoms, etc. Therefore, the plasma chemistry is not based on the Arrhenius equation with temperature dependencies of the rate of chemical reactions r(T) given ony by one common temperature T (in Kelvins) of reactants in the chemical reactor:

$$r\left(T\right) \approx \exp\left(-E_a / kT\right) \tag{2.30}$$

where $E_a$ is the activation energy of a given chemical reaction and k (or $k_B$) is the Boltzman constant. Instead of one temperature T, the low-pressure plasmas in molecular gas can be characterized by six different temperatures and by electrons as the main energy transporters (distributors).

In weakly ionized molecular gas at reduced pressure, depending on the energy (temperature $T_e$) of the electrons, most of the gas molecules remain neutral and they can be forced to move only at the gas temperature $T_g$. We assume a *translation temperature* ($T_t$) roughly equal to the temperature of the gas ($T_g \approx T_t$). With energy increasing for the electrons colliding with the molecules, the atoms in the molecules can start to rotate at the *rotational temperature* and some pairs of the inner atoms in the molecule can vibrate and acquire a *vibrational temperature*. At higher energies of the incident electrons, the molecules can be excited and acquire an additional potential energy characterized by an *electronic temperature*, due to shifts of internal electrons in the valence shells of the atoms or molecules to the higher orbits. When the bombarding electrons have energies high enough for ionization, the molecule can lose an electron and acquire positive charge. The produced ions have a lower energy (temperature) than the electrons. Therefore, the plasma in a molecular gas can be characterized by several temperatures (or energies, when multiplied by the Boltzman constant):

- The gas temperature ($T_g$)
- Translational temperature ($T_t$)
- Rotational temperature ($T_r$)
- Vibrational temperature ($T_v$)
- Electronic temperature ($T_{el}$) for electrons at valence orbits
- Plasma ion temperature ($T_i$)
- Electron temperature ($T_e$)

Therefore, in the non-equilibrium plasmas in molecular gas, one can recognize the following:

$$T_g \approx T_t < T_r < T_v << T_{el} \text{ and } T_i << T_e \tag{2.31}$$

In atomic gases, no rotation and vibration temperatures can be considered. The electron temperature $T_e$ (or energy $W_e \approx kT_e$) in the gas discharges can often be tuned for possible reactions by reaching the threshold at the start of the desired reaction. For such tuning, one must know the generated electron energy or the electron energy distribution function (EEDF). The Langmuir probe can measure the *EEDF, which characterizes how many electrons have a certain energy throughout an entire energy scale*. For the equilibrium plasmas, the typical distribution is the Maxwell-Boltzmann one; in the chemically reactive plasmas at reduced gas pressures, it can differ from the Maxwellian one.

### 2.5.3 Microwave Generation of Plasmas

As explained in Section 2.3, all plasmas exhibit certain critical density ($n_{crit}$), above which the incident electromagnetic waves are strongly reflected from the plasma. This density depends on the plasma electron frequency ($\omega_{pe}$) and on the frequency square ($f^2$) of the incident wave, see Eqs. (2.3.5 and 2.3.6). It means that the microwave power can generate higher plasma density than lower frequency AC powers, like RF power. Indeed, the critical density for the microwave power ($f_{MW} = 2.4$ GHz) is $n_{crit} \approx 7 \times 10^{10}$ cm$^{-3}$; for a typical radio frequency power ($f_{RF} = 13.56$ MHz), it is only $n_{crit} \approx 2 \times 10^6$ cm$^{-3}$. For more detailed explanations, one can find main differences between microwave and RF plasmas introduced and explained in more details, see, e.g. in Ref. [2.23].

Without applying magnetic fields in the microwave plasma, one can expect the plasma densities of the order of $10^{10}$ cm$^{-3}$. The electron temperatures can reach several eV, and the temperatures of heavier ions can be lower due to higher inertia. The plasma parameters depend on the gas pressure, type of used gas, and used power launching arrangements.

The power launching in microwave generation of plasmas can be arranged in different systems using, e.g. special electrodes and surface wave launchers or by using electrodeless systems, resonators, etc. A comprehensive topical review on the low-pressure microwave plasma systems can be found, in Ref. [2.24]. Several microwave systems for processing applications, which will be described later, are illustrated in Figure 2.11. Typical gas pressures in these systems are between $10^{-2}$ Torr and 10 Torr. However, some of the systems shown in Figure 2.11 can be used for microwave generation of an atmospheric plasma. Atmospheric plasma systems and applications will be described in Chapter 5. In many practical applications, the microwave plasmas are generated in the ECR regimes using the magnetic field defined in Eq. (2.15). Plasma densities can reach up to $10^{13}$ cm$^{-3}$. Such densities are possible even at 1 mTorr pressure and lower, with the ionization degree reaching 100%.

Because of the high frequency of the microwave power, the motions of electrons and ions in the plasma are decoupled and the space-charge sheaths at surfaces immersed in the plasma are thin with capacitive character. The sheaths are thinner than the ion sheaths in the RF plasmas. In an example of a simple calculation, an electron displacement to distance "x" in an electric field $E \approx 1000$ V/m with the frequency $f = 2.4$ GHz gives the following:

$$x = eE / [m_e(2\pi f)^2] \approx 0.77\ \mu\text{m} \tag{2.32}$$

This is about two orders of magnitude shorter than the typical sheath thicknesses in the RF discharges presented by the graph shown in Figure 2.9.

The thickness of the microwave sheath can be estimated to be between 3 and 14 times of the Debye length, see Refs. [2.25–2.27]. For an electron density of

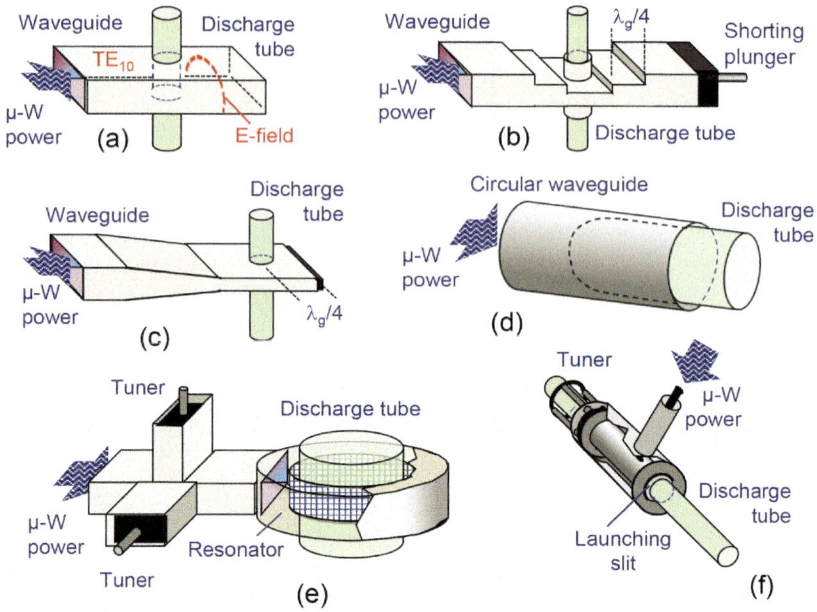


**Figure 2.11** Typical microwave systems for the plasma generation at reduced pressures. The discharge tubes are vacuum pumped with continuous delivery of a plasma gas (not shown). (a) Simple waveguide system (can be terminated by short or by dummy load). (b) Step tapered waveguide. (c) Tapered waveguide. (d) Discharge tube at the outlet of circular waveguide. (e) Microwave resonator. (f) Surfatron launcher (coaxial type).

$10^{10}$ $cm^{-3}$ and an electron energy of 5 eV, the value calculated from Eqs. (2.3 and 2.4) gives the Debye length of about $\lambda_D \approx 0.2$ mm and the sheath thickness $d_s$ between 0.6 mm and 2.8 mm, The Monte Carlo simulations particle-in-cell (PIC) confirmed that the sheath thickness depends inversely on the generator frequency, see Ref. [2.28]:

$$d_s \sim (1/\omega)^{0.87} \text{ and roughly } d_s \approx p^{-1/2}\,\omega^{-1} \tag{2.33}$$

These expressions for the microwave systems with frequency $\omega = 2\pi$ 2.4 GHz $\approx$ 15 GHz confirm thin sheath thickness. Indeed, electrodes immersed in the microwave discharge exhibit almost no visible sheaths similar to those observed in the RF discharges at three orders of lower frequencies. Recent measurements of the sheath thickness at frequencies up to 3 GHz confirmed sheath thickness below 0.4 mm, see Ref. [2.29].

The thickness of the space-charge sheath and the energy of the ions bombarding the electrode decreases with the increasing frequency. In the microwave range, the sheath resistivity is high, and the continuity of the current flow is based on the displacement currents through the sheath. In this case, the role of electrodes in the generation of charged particles is negligible because γ-processes associated with the secondary electron emission from the electrodes can be neglected. The discharge exists in the form of an α-discharge, in which the volume ionization is of main importance. Due to the low energy of ions, the microwave plasma is inefficient in ion-based processing, like sputtering or ion etching. Such purposes can be fulfilled only by combinations of the microwave power with other kinds of power sources, e.g. RF or DC biasing of substrates. On the other hand, the high electron density and possible control of the electron energies can be favorable in many plasma-chemical processes when forming *reactive radicals*, of which the *radical* represents part of a dissociated molecule. The radicals are unstable and, therefore, chemically reactive. The chemical reactions in the plasma must be based only between heavy particles (ions, atoms, molecules). The important role of electrons here is to activate these particles. Only heavy particles have proper mass for chemical reactions and they bear a certain chemical state needed for different reactions. The collisions with electrons can change the energy (state) of such particles and form reactive radicals, which can provide several orders higher reaction rates than those in the conventional equilibrium chemical reactions.

Figure 2.12 illustrates the comparison of optical spectra from nitrogen plasma generated at comparable conditions by the microwave power (2.35

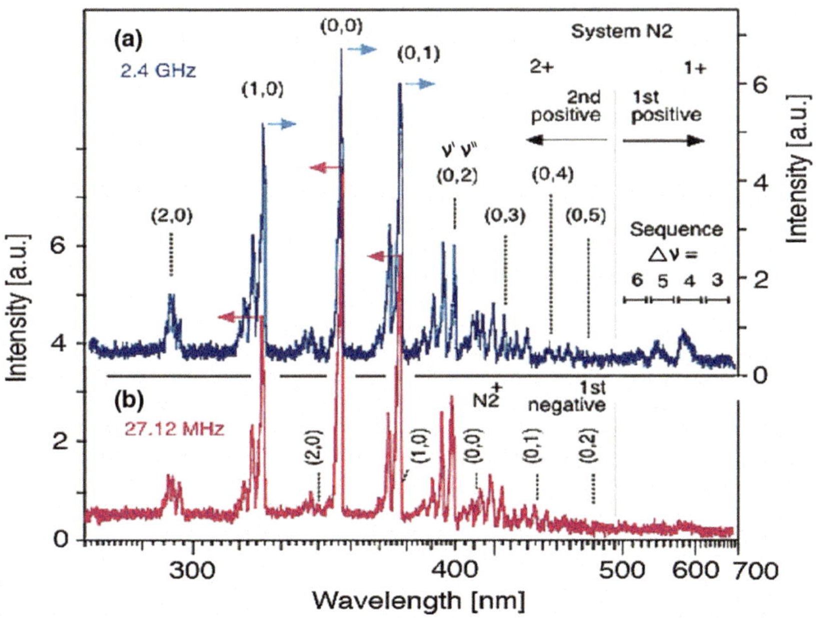


**Figure 2.12** Optical emission spectra from nitrogen plasma generated by the microwave power of 15 W (a) and an RF power of 40 W. The pressure was kept at 0.5 Torr (67 Pa).

GHz, 15 W) and the RF power (40 W, 27.12 MHz). Despite lower power, the first positive system ($1^+N_2$) and most band heads have higher intensity in the microwave discharge than in the RF discharge, see Ref. [2.30]. The Surfatron launcher used for generation of this nitrogen plasma is shown in Figure 2.11f.

Interactions of high-energy electrons with the gas particles can produce reactive atoms and radicals, which can generate chemical reactions not possible in the conventional equilibrium chemical systems. Activity of such plasmas can be used in different plasma-chemical applications, see e.g. Refs. [2.31, 2.32]. The ion energies in the microwave plasma at reduced or low pressures can be efficiently controlled by auxiliary electric fields and magnetic fields to allow, e.g. an ion bombardment for sputtering of target cathodes or the production of desirable metastable particles for alternative way of deposition of films onto substrates positioned far from the plasma region. Principles and some special devices for such applications will be described in the following parts of this book.

## References


[2.1] J.S. Townsend and H.T. Tizard: "The motion of electrons in gases", Proceedings of the Royal Society, June 2 (1913) 336–347.

[2.2] J.S. Townsend: "Electricity in gases", Clarendon Press, Oxford, 1915.

[2.3] S.C. Brown: "Basic data of plasma physics", M.I.T. Press, 1966.

[2.4] Y.P. Raizer: "Gas discharge physics", Springer-Verlag, Berlin, Heidelberg, 1991.

[2.5] F. Paschen: "Ueber die zum Funkenübergang in Luft, Wasserstoff und Kohlensäure bei verschiedenen Drucken erforderliche Potentialdifferenz"; Ann. Phys. **273** (5) (1889) 69–75.

[2.6] F.F. Chen: "Introduction to plasma physics", Plenum Press, New York, 1974.

[2.7] F.F. Chen: "Helicon plasma sources", Chapter 1 in O.A. Popov, ed.: "High density plasma sources", Noyes Publications, Park Ridge, New Jersey, 1995.

[2.8] R.W. Boswell: "Plasma production using a standing helicon wave"; Phys. Lett. **A33** (1970) 457.

[2.9] F.F. Chen: "Experiments on helicon plasma sources"; J. Vac. Sci. Technol. **A10** (1992) 1389.

[2.10] T. Mieno, T. Shoji, and K. Kadota: "Control of Hydrocarbon Radicals and Amorphous Carbon Film Deposition by Means of RF Whistler Wave Discharge"; Jpn. J. Appl. Phys. **31** (1992) 1879.

[2.11] D.L. Flamm: "Frequency effects in plasma etching"; J. Vac. Sci. Technol. **A4** (3) (1986) 729.

[2.12] S. Kakuta, T. Makabe, and F. Tochikubo: "Frequency dependence on the structure of radio frequency glow discharges in Ar"; J. Appl. Phys. **74** (8) (1993) 4907.

[2.13] J. Perrin, P.R. i Cabarrocas, B. Allain, and J.-M. Friedt: "a-Si:H Deposition from SiH4 and Si2H6 rf-Discharges: Pressure and Temperature Dependence of Film Growth in Relation to α-γ Discharge Transition"; Jap. J. Appl. Phys. **27** (11) (1988) 2041.

[2.14] K.-U. Riemann: "Theoretical analysis of the electrode sheath in rf discharges"; J. Appl. Phys. **65** (3) (1989) 999.

[2.15] M.D. Gill: "Sustaining mechanisms in rf plasmas"; Vacuum **34** (1984) 357.

[2.16] D. Vender and R.W. Boswell: "Electron-sheathinteraction in capacitive radio-frequency plasmas"; J. Vac. Sci. Technol. **A10** (1992) 1331.

[2.17] H.R. Koenig and L.I. Maissel: "Application of rfdischarges to sputtering"; IBM J. Res. Dev. **14** (1970) 168.

[2.18] J.W. Coburn and E. Kay: "Positive ionbombardment of substrates in rf diode glow discharge sputtering"; J. Appl. Phys. **43** (1972) 4965.

[2.19] Y. Wang: "Practical definition of the plasma sheath edge for modeling planar glow discharges"; Appl. Phys. Lett. **66** (18) (1995) 2329.

[2.20] M. Mutsukura, K. Kobayashi, and Y. Machi: "Plasma sheath thickness in radio-frequency discharges"; J. Appl. Phys. **68** (1990) 2657.

[2.21] J. Wild, P. Kudrna, M. Tichy, V. Nevrly, M. Strizik, P. Bitala, B. Filipi, and Z. Zelinger: "Electron temperature measurement in a premixed flat flame using the double probe method"; Contrib. Plasma Phys. **52** (8) (2012) 692–698. http://dx.doi.org/10.1002/ctpp.201200005.

[2.22] R.C. West, M.J. Astle, and W.H. Beyer, eds.: "CRC handbook of chemistry and physics", 64th edition, CRC Press Inc., Boca Raton, Florida, 1983–1984. ISBN-0-8493-0464-4. Section F, page 163 "Atmospheric Electricity" by Hans Dolezalek.

[2.23] R.G. Bosisio, M.R. Wertheimer, and C.F. Weissfloch: "Generation of large volume microwave plasmas"; J. Phys. E **6** (1973) 628–630.

[2.24] Y.A. Lebedev: "Microwave discharges at low pressures and peculiarities of the processes in strongly non-uniform plasma"; Plasma Sources Sci. Technol. **24** (2015) 053001 (39pp). http://dx.doi.org/10.1088/0963-0252/24/5/053001.

[2.25] V. Godyak: "Comments on plasma diagnostics with microwave probes"; Phys. Plasmas **24** (2017) 060702 (4 pp).

[2.26] J. Oberrath and R.P. Brinkmann: "Active plasma resonance spectroscopy: a kinetic functional analytic description"; Plasma Sources Sci. Technol. **23** (2014) 045006 (14 pp).

[2.27] P. Chabert: "What is the size of a floating sheath?"; Plasma Sources Sci. Technol. **23** (2014) 065042 (4 pp).

[2.28] M. Surendra and D.B. Graves: “Capacitively coupled glow discharges at frequencies above 13.56 MHz”; Appl. Phys. Lett. **59** (17) (1991) 2091.https://doi.org/10.1063/1.106112.

[2.29] A. Khomenko and S. Macheret: “Diagnostics of small plasma discharges using probingin wide range of microwave frequencies”; Appl. Phys. Lett. 116, 023501 (2020); doi:10.1063/1.5135023.

[2.30] L. Bardos, J. Musil, and P. Taras: “Differences between microwave and RF activation of nitrogen for the P ECVD process”; J. Phys. D: Appl. Phys. **15** (1982) L79.

[2.31] F.K. McTaggart: “Plasma chemistry in electrical discharges”, Elsevier, Amsterdam, 1967.

[2.32] L.S. Polak and Y.A. Lebedev, eds.: “Plasma chemistry”, Cambridge International Science Publishing, London, 1998. ISBN 1898326223, 9781898326229.

# 3

# Interactions of Plasmas with Solids and Gases

We have used so far the term *plasma processing* without more detailed explanations or examples. In Chapter 2, a non-equilibrium (cold or non-thermal) gas discharge plasma has been described and introduced as a suitable medium for processing treatments of solid substrates or gases at temperatures well below 1000 °C. We will describe how and in what applications such cold plasma can be used and what results can be expected after such plasma processing treatments. For more details and theoretical explanations in this field, use fundamental books, see, e.g. Ref. [3.1]. The plasma processing will be described first at *reduced gas pressures* in an interval between 1.5 Pa and 15 MPa (≈ 0.01 Torr and 100 Torr) or at *low gas pressures* ≤ 1.5 Pa (≤ 0.01 Torr). In Chapter 5, *the atmospheric and higher gas pressures* will be introduced and described together with the novel cold atmospheric plasma (CAP) sources powered by microwaves.

## 3.1 Plasma Processing, PVD, and PE CVD

The cold gas discharge plasmas can be used in almost all types of the surface treatments. In many cases, such processing has been used as an industrial technology. The plasma processing can be considered as a particular part of the science and technology in *Plasma Science.* It is based on interactions of the plasma with nearly all kinds of materials, including solids, gases, and liquids. For a long time, plasma interactions with liquids have been typical rather for hot plasmas used in the plasma metallurgy. However, in the recent decades, the cold plasma at atmospheric pressure has been found efficient in treatments of different liquids, and it has been applied on liquid surfaces or submersed inside the liquid. Some experiments and devices will be described in Chapter 5, and a concise review on microwave plasma systems in liquids can be found in Ref. [3.2].

The plasma interactions with gases are often used in the plasma chemistry for changes in the chemical composition of gases, gas cleaning from effluents and poisons, gas activations, etc. However, for the plasma processing at reduced

*Microwave Plasma Sources and Methods in Processing Technology*, First Edition.
Ladislav Bárdoš and Hana Baránková.

and low pressures, the most typical and most widespread applications are interactions of the plasma with solid substrate surfaces. Such treatments are typical mainly in the microelectronic and micromechanic applications, where they enable controlled changes of the surface properties, e.g. its geometry, chemical composition, hardness, structure, morphology. The results depend on the type and parameters of the plasma used and on the solid surface.

Three basic types of surface processing exist:

1) Deposition of films
2) Etching or cleaning of the surface
3) Change of the chemical composition (e.g. an oxidation) and chemical-mechanical properties (corrosion resistance and/or hardness by surface nitriding or carburizing)

Typical processes in the plasma treatment of solid substrates can be explained using Figure 3.1. The sample to be processed is placed in a vacuum chamber (reactor), with steady pumping and simultaneous inflows of the gaseous components (e.g., gases A, B, C, ...) necessary for the desired plasma-chemical processes.

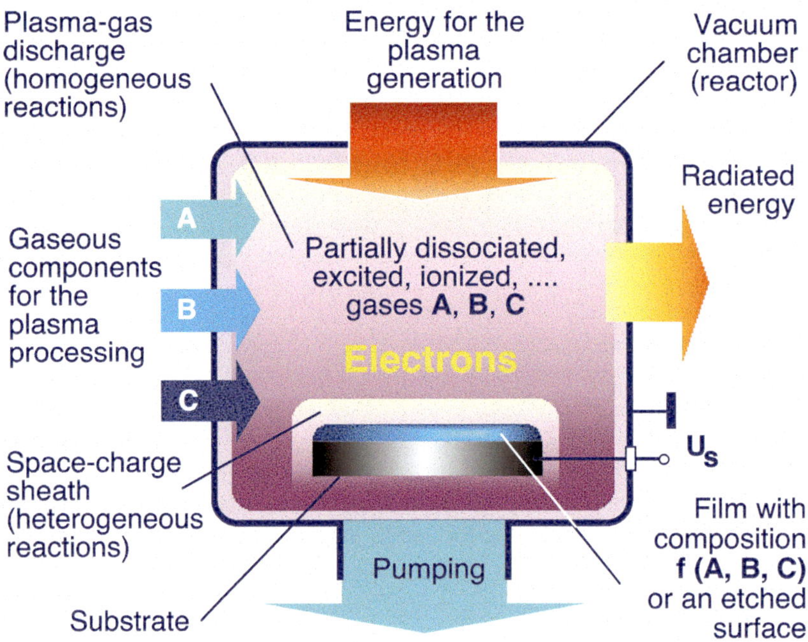


**Figure 3.1** Schematic representation of the particle and energy interactions in the reduced or low-pressure plasma in a reactor, with a solid substrate placed inside the reactor and biased by auxiliary voltage ($U_S$) with respect to the grounded reactor walls.

The involved gas mixture is often noted as the *working gas*. For example, in the processing of the deposition of a thin film on a sample surface, the chemical composition of growing the film will generally be a function f (A, B, C, ...) of gases involved in the working gas mixture. The ratio of the inflowing amounts of the respective gases corresponds to the ratio of their *partial pressures* ($p_A$, $p_B$, $p_C$), and they define the density of particles of these gases in the reactor. The sum of the partial pressures of the components A, B, and C gives the total gas pressure ($p_{TOT}$) in the reactor:

$$p_{TOT} = p_A + p_B + p_C + \dots \quad (3.1)$$

The *pumping speed* must be stable to keep the selected partial gas pressures in the reactor constant at the selected inflow rates of individual gas components. The available vacuum technology for such purposes, including different kinds of pumps and measuring sensors, is described well in the literature, see, e.g. Ref. [3.3]. We will refer only to the mechanical pumping for the reduced pressure processing (the rotary pumps and the Roots blowers) and to the high-vacuum pumps (e.g. the turbomolecular pumps, oil diffusion pumps, or cryogenic pumps) for the high-vacuum processing. The high-vacuum pumping always requires the high-vacuum pumps and the mechanical pumps for the backing of the high-vacuum pumps and for the pre-pumping down the reactor chambers from atmospheric pressure after closing all openings.

When the low-pressure gas discharge plasma is generated in the reactor by an electromagnetic power, the neutral particles in the working gas are partially ionized, excited, or dissociated in cases when a molecular gas is present. Depending on the generating parameters, e.g. the total pressure, parameters of the generator (frequency and power), types of the gases used in the working mixture, the plasma density (n) is represented by the density of positive ions and by an equal number of electrons (or a sum of electrons and negative ions in case of an electro-negative gas). In typical processing systems, the plasma density (n) reaches the order of the following:

$$n = n_e + \left(n_i^-\right) = n_i^+ \approx 10^8 - 10^{11} \text{cm}^{-3} \quad (3.2)$$

The density of neutral molecules (or atoms) ($n_{gas}$) in the working gas has been defined by Eq. (2.26) in Chapter 2. However, for rough estimates in the processing applications, it can be calculated as the following:

$$n_{gas}\left[\text{cm}^{-3}\right] \approx 10^{14} p_{TOT}\left[\text{Pa}\right] \approx 10^{16} p_{TOT}\left[\text{Torr}\right] \quad (3.3)$$

The degree of ionization $\delta_i = n/n_{gas}$ defined in Eq. (2.25) is usually less than 0.01 (1%) in the discharges for plasma processing at pressures 1–100 Pa. At low pressures, $p_{TOT} < 1$ Pa; the degree of ionization can be higher, (≥ 10%), particularly in the microwave plasma in a magnetic confinement or in the resonant generation regimes, e.g. in the microwave electron cyclotron resonance (ECR) systems, Refs. [3.4, 3.5], where we speak about the high-density plasma. In such plasmas, the electron density is usually $n_e \geq 10^{12}\ \mathrm{cm}^{-3}$.

The particle energies in processing plasmas at gas pressures between 10 and 100 Pa do not usually reach high values. An average energy of electrons usually does not exceed 10 eV (i.e. $T_e \leq 10^6$ K), which is the order of typical ionization potentials as discussed in Chapter 2. The energy of heavier ions is lower, roughly $T_i < 0.1\ T_e$. As discussed (Section 2.1, Chapter 2), all values found in the plasma can differ substantially at surfaces separated by the space-charge sheaths. An electric field across the space-charge sheath can accelerate ions even up to several tens of eV, see, e.g. Ref. [3.6]. Therefore, the space-charge sheath almost always represents an important transition zone between the processes (reactions) in the volume and bulk of plasma (homogeneous reactions) and the surface of the solid substrate (heterogeneous reactions), see scheme in Figure 3.2.

As can be understood from Figure 3.2, the general reaction scheme in the plasma processing of the surface of a substrate in chemically reactive plasma can be written in the following way:

$$\begin{array}{lll} \text{homogeneous plasma} - \text{chem.} \Leftrightarrow & \text{heterogeneous reactions} \Leftrightarrow & \text{solid film or} \\ \textit{reactions}\ (\text{in the gas phase}) & (\text{at the gas} - \text{solid interface}) & \text{etched surface} \end{array} \tag{3.4}$$

Neutral gas components introduced into the plasma are transformed by the homogeneous (volume) reactions to active forms, e.g. radicals, excited or ionized atoms. These species collide with the surface and with other particles.

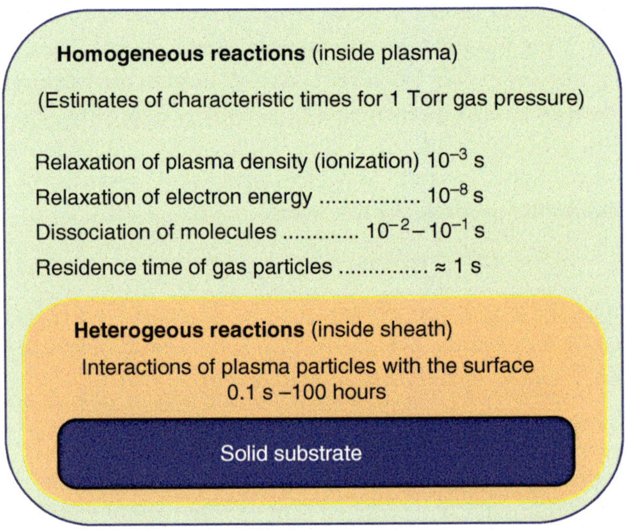


**Figure 3.2** The homogeneous and heterogeneous reactions in the plasma processing. Time estimates for typical processes at 1 Torr (133 Pa) gas pressure are taken from Ref. [3.7].

The resulting product of such collisions can be a new kind of gas, transformed from original gases, or a solid deposit coated on the substrate surface in form of a thin film (coating). The growing film as well as the original substrate surface can be simultaneously etched (gasified) by the activated reactive particles or sputtered by an ion bombardment. The homogeneous reactions can then be affected by their own products on the substrate surface. Therefore, the properties of the substrate, e.g. the chemical composition, externally applied bias, or geometry, play an important role in the processing technology.

The plasma processing of surfaces of solid substrates can be characterized as follows: *the plasma processing is an interaction of the gas discharge plasma or its products with the surface of substrate, which results in changes of the surface properties and/or its geometry.*

The deposition of films onto surfaces of solid substrates can be realized by different methods. The selection of suitable method depends on the desired application. It is always necessary to select a proper deposition technology according to the following:

- The substrate to be used defined by chemical composition of the substrate surface, temperature limits (melting, sublimation, etc.), geometry of the substrate, etc.
- The film to be deposited defined by desired chemical composition, desired thickness and its uniformity or its distribution over the surface, desired mechanical, chemical, optical, electrical, and other film properties

The industrial applications of the particular deposition technology must account for an economy of the selected processes, e.g. the productivity, costs, availability. Therefore, in many cases, finding satisfactory technology for the desired application is complicated.

Two groups of the film deposition technologies are available, according to their respective principles: (1) Chemical Vapor Deposition (CVD) and (2) Physical Vapor Deposition (PVD).

Typical representatives of group (1) can be chemical precursors (e.g. metal chlorides, iodides) used in the deposition of metal films at elevated temperature. Typical representatives of group (2) can be the deposition of metals using thermal evaporation of selected metal in the high vacuum. In plasma processing, the physical processes and plasma chemistry play important roles. This can be demonstrated in the *Plasma-Enhanced Chemical Vapor Deposition* (PE CVD). Other notations with the equivalent meaning are frequently used: Plasma-Activated CVD or Plasma-Assisted CVD (PA CVD), and Plasma-Induced CVD (PI CVD). The PE CVD methods are based on physical-chemical principles of the generation of the discharge and simultaneous formation of the chemically active species. In some gas mixtures, the discharge can induce fast plasma-chemical reactions at low gas temperatures, which cannot exist in classical chemistry. Examples of the PE CVD processes will be introduced in Chapters 4, 5, and 6.

The gaseous components A, B, and C described in Figure 3.1 could be represented by the vapor of an evaporated metal or by particles sputtered from some metal target by ion bombardment (e.g. in the magnetron sputtering of a cathode target). The reaction principles introduced in Figure 3.1 have general validity. The reactive sputtering and reactive evaporation in the film deposition are assumed to be the PVD methods, where the sputtered and evaporated material reacts with the reactive gas plasma. Such methods belong also to the general PE CVD (or PA CVD) methods. The deposition of any particular film by the PE CVD has usually several possibilities based on the choice of the working gas. On the other side, the change in the chemical composition of the working gas will change the composition of the deposited film. This possibility opens a number of technological variations in the film growth and the resulting film properties. Some of such cases will be described in Chapter 4.

Contrary to the plasma-based depositions in the plasma-based etching (also called *dry etching*), the plasma produces chemically reactive particles, which react with the solid surface and form volatile compounds. These compounds are pumped out from the reactor, and in a continuous process, the considerable parts of the surface can be etched away. The dry etching methods depend on the proper choice of the working gas. The etching gas must produce species, which can react chemically with the material to be etched, and the key requirement is the volatility of the etched product, see Ref. [3.8]. Etching mechanisms differ in different gases and in plasma parameters. At least three different etching mechanisms exist, see Ref. [3.9]:

1) Sputtering-based (based on a surface erosion by an ion bombardment)
2) Chemical etching (based on chemical reactions on the surface)
3) Ion-enhanced chemical etching (also RIE – the reactive ion etching)

The sputtering process is important for the film deposition and it can assist also in the dry etching processes. In real processes, both mechanisms occur, and the resulting effect depends on the ratio of the rates of involved processes. The chemical compositions of typical gases using in dry etching are usually based on Fluorine (F), Chlorine (Cl), Oxygen (O), and Hydrogen (H).

## 3.2 Sputtering, Evaporation, Dry Etching, Cleaning, and Oxidation of Surfaces

The low-pressure interactions of the charged particles from plasma with solid surfaces depend on the particle energy in the plasma and the surface bias with respect to the plasma. Moreover, because of the difference in their mass, ions and electrons differ substantially when they interact with the solid surfaces. Accelerated ions can bring substantial mechanical momentum to the surface, and they can interact with individual surface atoms, transfer their momentum, and eventually release-sputter the atoms from the surface. Instead, the light electrons interact with the oscillating crystal lattice in the surface (also called phonon interactions), and above certain levels of electron flux density and energy in the electrons, they can increase the substrate temperature up to its evaporation.

The electron current in metals is known to cause ohmic heating, which depends on the metal resistance and the current level. In a simplified approach, such ohmic heating may be described as a motion of an electron cloud interacting with electrons in the valence shells of individual metal atoms causing enhanced vibrations of the metal lattice, i.e., the heating. Similarly, the flux of electrons interacting with a positively biased surface interacts with the surface atoms, enhances the lattice vibrations, and heats the surface. Depending on the thermal conductivity of the substrate, the heat from the surface can be distributed with a certain gradient to the whole substrate. Therefore, in the thermally insulated systems, the heating by an electron beam can be used for melting positively biased targets and their *vacuum evaporation*. This is usually applied in high vacuum chambers to avoid interactions of the produced vapor with residual gases. The electron beams about 10–100 $A/cm^2$ can melt most metals at bias voltages of about 10 V. Indeed, the power delivered by this beam is as high as 0.1–1 $kW/cm^2$. The melting temperature of metals and their saturated (maximum) vapor pressure differ in individual metals, see data, e.g. Ref. [3.10]. Certain metals exhibit a high pressure of their own metal vapor even below their melting point. The presence of metal vapor from electrodes can affect the plasma parameters in the system. Moreover, electrons with low energy (several eV up to about 30 eV) can attach to the particles of the electro-negative gas and form negative ions. A typical example is oxygen, where the negative oxygen ions can oxidize solid substrates. These processes are called the *plasma oxidation* or *plasma anodic oxidation*. An example of the plasma anodic oxidation of silicon is shown in Figure 3.3. More detailed experimental data will be introduced and discussed later in Chapter 4.

An ion bombardment of surfaces can generate heat. However, due to the large ion mass compared to the electrons, the ion interactions with electrode surfaces have a substantially different character, usually more aggressive and destructive. The bombardment of the negative electrode surfaces by certain density of positive ions under negative acceleration voltages can cause numerous effects depending on ion energy (affected by the bias), ion mass, angle of the incidence, and surface conditions, like roughness, temperature, chemical composition, etc. Processes on the cathodic and anodic surfaces are schematically depicted in Figure 3.3. The typical processes are the following:

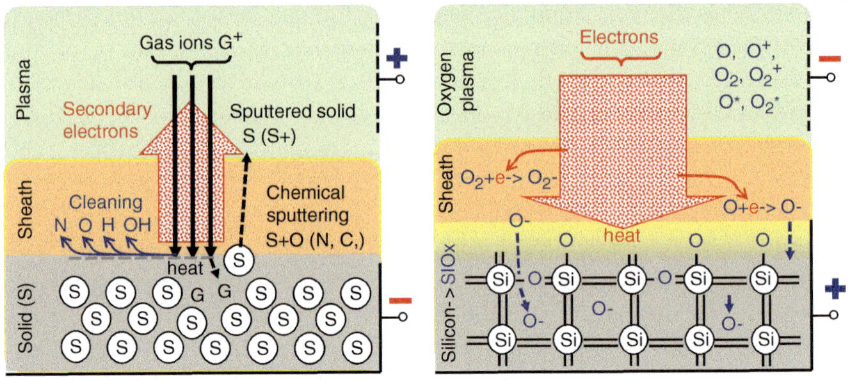


**Figure 3.3** Schematic representation of possible processes at negative (left) and positive (right) biased substrates. An incident ion is considered to be a gas ion (G +). The electrode target is considered to be metallic and consist of metal atoms (S). An example of the plasma anodic oxidation shows silicon oxidation based on a diffusion of negative oxygen ions.

- Cleaning the cathode surface (at ion energies of about 0.1 eV–10 eV) when weakly bonded impurities are loosened by the incident ion and forced to leave the surface.
- Sputtering of the cathode surface above a certain threshold energy of ions depending on the ion and the surface (at typical ion energies for sputtering, about 50 eV–1 keV), when the surface atoms are loosened by an ion impact.
- Anodic oxidation of an anode surface by low-energy electrons (roughly below 50 eV), with possible heating and partial evaporation by higher energy electrons.

With a large mechanical momentum, the ions can cause surface bonds to break and surface particles to loose (typically atoms) from the cathode surface (often called a target). Such ion-surface interactions caused by *the momentum transfer* are denoted as a *sputtering*. Because of the short characteristic time of the momentum transfer from ions to the surface atoms, there is only limited yield of energy transferred from ions to the oscillations of crystal lattice of the solid (ion-phonon interactions heating the surface). Therefore, the typical feature of the sputtering process is an absence of high surface temperatures. However, if the bombarded surface during sputtering is thermally insulated and not cooled, its temperature can easily increase, similar to what occurs in electron bombardment. For materials like aluminum (Al), zinc (Zn), and tin (Sn), the temperature can quickly reach the melting point. At this stage, the contribution of the sputtering mechanisms on the total production of particles from the target can be overbalanced by the contribution of evaporation.

Therefore, in the most of practical cases, the cathode (target) is cooled during sputtering to suppress heating of the target and to avoid its melting.

Important characteristic values in the sputtering are the *sputtering threshold* and the *sputtering yield*. The sputtering threshold is a minimum ion energy, necessary for the sputtering. Typical threshold values are around 50 eV. This threshold depends on the energy of surface bonds on the target. The sputtering yield represents the number of the target atoms produced by one incident ion hitting the target during sputtering. With identical conditions, the sputtering yields differ for different cathode target metals, see, e.g. Ref. [3.11]. However, the value of the sputtering yield is not constant. On the contrary, for a given target metal, the yield depends on many factors, the most important being the incident ion energy and angle of incidence.

The *microwave plasma* is not primarily suitable for sputtering although it is possible to perform sputtering with help of an auxiliary bias applied to the target. Moreover, the liberation of particles from an ion-bombarded target can be performed also at ion energies lower than the sputtering threshold by using *chemical sputtering*. With chemical sputtering, the incident ions initiate suitable surface chemical reactions leading to the liberation of the surface atoms or molecules. Such processes can be considered a kind of *dry etching* of the surface. However, while dry etching is occurring, the surface chemical reactions result in a volatile product containing the surface atoms to be pumped out; in the chemical sputtering, the produced atoms or compounds are supposed to be deposited as films on the selected surfaces. The plasma etching (dry etching) of surfaces is usually a faster and better controllable process than chemical etching. Details of this technology are available in several publications, e.g. Refs. [3.8, 3.9].

## 3.3 Particle Transport in Plasma Processing and Effects of Gas Pressure

A difference between neutral particles in the plasma (e.g. atoms, molecules excited neutrals, radicals) and the charged particles (ions and electrons) is an easier control of the motions of charged particles by electric and magnetic fields. To affect and control the transport of neutral particles is always more complicated. However, because particle types are usually important in the plasma processing, and in many applications, both particle types are to be transported from the plasma generation area (the plasma source) to the substrate; the individual control methods will be shortly described in Section 3.3.1. The transport will be considered as the driving of selected particles through certain transport region between the plasma source and the substrate.

### 3.3.1 Movements of Neutral Particles

The electromagnetic fields do not affect the neutral particles (mass $m_{gas}$) in the plasma. Without any forcing power, they move stochastically (with random motions in all directions) with an average thermal velocity ($v_{gas}$). In an idealized approach, when the particles are considered as independent hard spheres, the thermal velocity depends only on the gas temperature ($T_{gas}$) and one can write the following:

$$m_{gas}v_{gas}^2/2 = 3\ k_B T_{gas}/2 \text{ or } v_{gas} = \left(3\ k_B T_{gas}/m_{gas}\right)^{1/2} \tag{3.5}$$

where $k_B$ is the Boltzmann constant. Using Eq. (3.5), the thermal velocity of argon (Ar) particles (using $m_{gas}$ = mass of proton $1.67\ 10^{-27}$ kg multiplied by the Ar mass number 39.95 = $6.67\ 10^{-26}$ kg) at room temperature ($T_{gas}$ = 300 K) should give $v_{gas} \approx 430$ m/s. The corresponding value for neutral (molecular) nitrogen ($m_{gas} = 2.34\ 10^{-26}$ kg) is as high as $v_{gas} \approx 720$ m/s. These values are surprisingly high and both are exceeding the sonic velocity (about 340 m/s at 20°C = 293 K). This surprise can be explained by the zero collisions considered in Eq. (3.5). Moreover, the particles are moving stochastically in all directions and all particles are supposed to have the same velocity. In a more realistic approach, it is necessary to consider that particles collide with each other and they can acquire different velocities over a broad interval of values. According to the kinetic theory of gases, the distribution of the particles as a function of their velocities obeys the Maxwell–Boltzmann distribution function:

$$\frac{dn_{gas}}{dv} = \frac{2n_{gas}}{\sqrt{\pi}}\left(\frac{m_{gas}}{2k_B T_{gas}}\right)^{\frac{3}{2}} v^2 \exp\left(-\frac{v^2 m_{gas}}{2k_B T_{gas}}\right) \tag{3.6}$$

Here $n_{gas} = n_{gas}$ (v) represents the number of gas particles in a unit volume having velocity v. The sum (integral) of particles over all possible velocities must be equal to the total gas particle density n present in the considered (unit) volume:

$$n = \int_0^\infty n_{gas}(v)dv \tag{3.7}$$

The most probable velocity in the *distribution function* given in Eq. (3.6), ascribed to the largest number of particles, is lower than values calculated from Eq. (3.5) but still high. For example, for nitrogen (N), it is about 400 m/s, see, e.g. Ref. [3.3].

In the description given above, the motions of neutral particles were considered in a uniform distribution of their density in the studied (unit) volume. This might cause an unrealistic impression that the movements of neutral particles cannot be controlled, but several possibilities exist as how to drive neutral particles. A simplest force driving the neutral particles is based on the gradients of their density. Particles with the space-dependent density n (x,y,z) are forced to diffuse from the regions with higher density to regions with lower density by the force vector $\mathbf{F}_D$:

$$\mathbf{F}_D \approx \text{grad } n(x, y, z) \tag{3.8}$$

In a one-dimensional approximation (e.g. along the x- axis), the total flux Φ of neutral particles caused by this force is expressed by the following:

$$\Phi = D\, n(x)/dx \tag{3.9}$$

D is the *diffusion coefficient*, characteristic for the given gas. The equation Eq. (3.9) is a one-dimensional expression of Fick's law (Φ = D grad n (x, y, z)). Fick's law is valid also for charged particles. All diffusion-based mechanisms are generally important in the of plasma processing.

Another type of the force affecting the movements of neutral particles can be understood on the basis of the general law for an ideal gas:

$$p_{gas} \cdot V_{gas} = N_{gas} k_B T_{gas} \tag{3.10}$$

$N_{gas} = n_{gas} \cdot V_{gas}$ represents the total number of particles in a considered gas volume ($V_{gas}$). The driving forces can be generated, e.g. by the temperature gradients or pressure gradients. The pressure gradients, created, e.g. by forcing gas through different gas nozzles, are often of particular importance in practical applications because they can enhance transport of the particles at high velocities.

### 3.3.2 Movements of Charged Particles

As noted above in Section 3.3.1, in contrast to the movements of neutral gas particles, the movements of electrons and ions can be controlled by electric and magnetic fields. The Lorentz force vector ($\mathbf{F}_{e,i}$), which affects motions of charged particles (electrons or ions), has an electric and a magnetic component. The electric component is linearly dependent on the intensity vector of the electric field (**E**), and the magnetic component is defined by the vector product between the particle velocity $\mathbf{v}_{e,i}$ (electrons or ions) and the vector of magnetic induction **B**:

$$\mathbf{F}_{e,i} = q_{e,i}\mathbf{E} + q_{e,i}\left(\mathbf{v}_{e,i} \times \mathbf{B}\right)\left[\text{N, C, m.s}^{-1}\text{, T}\right] \tag{3.11}$$

Here, $q_{e,i}$ denotes the particle charge (electron or ion). The single ionized ion has an opposite charge to an electron with the same absolute value (in coulombs [C] = [A s]):

$$|q_i| = |q_e| = e = 1.6\ \text{x}10^{-19}\,\text{A s} \tag{3.12}$$

The magnetic component of the Lorentz force vector acts only on those electrons and ions which have a non-zero vector product (**v** x **B**), i.e. those having a non-zero component of the velocity $v_{norm\,(e,i)}$ normal to the vector **B**. Particles moving parallel with vector **B** are not affected, yet other electrons and ions (having the mass $m_{e,i}$) are forced to gyrate around the magnetic power lines at the *Larmor radii* $r_L$ defined as the following:

$$r_L = m_{e,i} v_{norm\,(e,i)} / \left( \left| q_{e,i} \right| B \right) \tag{3.13}$$

Here, $|q_{e,i}|$ is an absolute (always positive) value of the charge and B is the value of the induction vector. The Larmor radius depends on the particle mass, and differs for electrons and for ions. Using the magnetic components of the Lorentz force vector, it is possible to shape special "magnetic mirrors" or vessels for the confinement of the charged particles in a selected space. Moreover, a non-uniform magnetic field with a certain gradient of the induction B composes an additional driving force on the charged particles. Such force can be expressed in the following way:

$$\mathbf{F_B} \approx \text{grad } B(x,y,z) \tag{3.14}$$

For example, it can be utilized in special magnetic traps for charged particles, i.e. for the *magnetic confinement* of the plasma avoiding contacts with the reactor walls. Many magnetic confinements (plasma traps) using static magnetic fields were proposed for the confinement of glow discharges, see Ref. [3.12]. In applications of electromagnetic power, the magnetic arrangements optimized with respect to the used electrodes can maximize the confinement of the plasma in desired area, reduce losses of charged particles outside the desired area, and allow plasma to sustain even at low gas pressures in the discharge chamber (even under $10^{-4}$ Torr ≈ 0.01 Pa).

Although the magnetic forces are useful for controlling the charged particles in plasmas, the most typical force for the direct driving of charged particles is the electric field, i.e. the electric component of the Lorentz force vector: $q_{e,I}$ **E** in Eq. (3.11). An important value in such particle motions is the mobility $\mu_p$ of particles (the units are $m^2/(V\ s)$), which is defined as the following:

$$\mu_p = v_d / E = q / \left( m \nu_m \right) \tag{3.15}$$

Here, $v_d$ is the drift velocity of the particle in the electric field with intensity E, $q = q_{e,I}$ is the charge (electron, ion), $m = m_{e,i}$ is the mass, and $\nu_m$ is the collision frequency with transfer of the particle momentum. Because the electron mass $m_e = 9.11 \times 10^{-31}$ kg is about three orders of magnitude smaller than the mass of ions ($m_i \geq 1.67 \times 10^{-27}$ kg = mass of proton), the mobility of the electrons and their corresponding ability to respond to the changes of the applied electromagnetic fields is higher than for ions. This capability can lead to higher electron energies than for ions (energy non-equilibrium). As discussed in Chapter 2, this is a characteristic feature of the low-pressure plasmas and an important property of plasmas powered by the high-frequency fields. One can conclude that the *electrons are the most important particles in the ionization processes in the gas discharges.*

A simple ionization path of a gas particle caused by the collision with an energetic electron (having energy exceeding the ionization threshold) can be expressed as the following:

$$\text{Energetic electron} + \text{Gas particle} \rightarrow \text{Positive Gas Ion} + 2\ \text{Electrons} \pm \text{Energy} \quad (3.16)$$

The Energy part in Eq. (3.16) is a term depending on the kind of gas. An excess energy is usually radiated, partly in an optical emission. Therefore, the optical emission spectroscopy (OES) is an important non-disturbing diagnostic tool in all gas discharge plasmas. If calibrated, the OES can identify the particles present in the plasma and their energies.

The Energetic electron in Eq. (3.16) is expected to have enough energy to ionize the gas. The energy of electrons is usually expressed in electron volts (eV), which is the energy acquired by an electron in an electric field along a potential difference of 1 volt. In Eq. (2.1) in Chapter 2, the energy of 1 eV is equivalent to the temperature of 11 600 K. To ionize the gas, the incident electrons must have a potential energy exceeding the minimum ionization potential ($V_{ioniz}$) of the given gas. Because of the collisions, neither electrons nor ions can simply acquire as much energy (velocity $v_{e,i}$) as would correspond to an entire potential drop along the unit length. The particle velocity in a stationary electric field (E) is expressed in terms accounting for the collisions with the neutral gas particles (the most probable collisions), represented by the collision frequency $\nu_{coll\,(e,i-gas)}$:

$$v_{e,i} = \frac{q_{e,i}}{m_{e,i}} \frac{E}{v_{coll(e,i\text{-}gas)}} \quad (3.17)$$

The role of collisions (and, therefore, the gas pressure) is important for the particle energy and for the overall character of the plasma. The motions of charged particles in the plasma are affected by the external electric, magnetic, or electromagnetic (oscillating) fields. As discussed, in the charged particle transport, the Coulomb forces are important, see Eq. (2.6) in Chapter 2.

## 3.3 Effect of the Gas Pressure on the Plasma Processing

As explained, an increasing gas pressure increases the collision frequency and decreases the mean-free paths of particles, see Section 2.5, Chapter 2, and Eqs. (2.27 and 2.28). This can affect the velocity of neutral and charged particles. An explicit direct effect of the collisions on the plasma processing systems exists.

The collisions during transport of the desired particles from their origin in the plasma generation region to the substrate surface can change their reactivity and chemical composition, their energy (velocity), and the directions of their motion. The effect of growing gas pressure is schematically illustrated in Figure 3.4.

Figure 3.4 shows three typical levels of the gas pressure

1) < 1 mTorr (< 0.13 Pa) with fewer than $10^{13}$ particles in $cm^3$
2) An interval 10 mTorr–10 Torr (1.3 Pa–1.33 × $10^3$ Pa) with $10^{14}$–$10^{17}$ particles/$cm^3$
3) > 10 Torr (> 1.33 × $10^3$ Pa) with more than $10^{17}$ particles/$cm^3$.

If the mean-free path, see Eq. (2.28) in Chapter 2, is longer than the transport (transmission) area, which depends on the reactor geometry and the position of substrates with respect to the plasma generation area (plasma source), then no collisions occur and the particles can reach the substrate in the state they were generated. This case is often denoted as *the line-of-site arrangement*. At

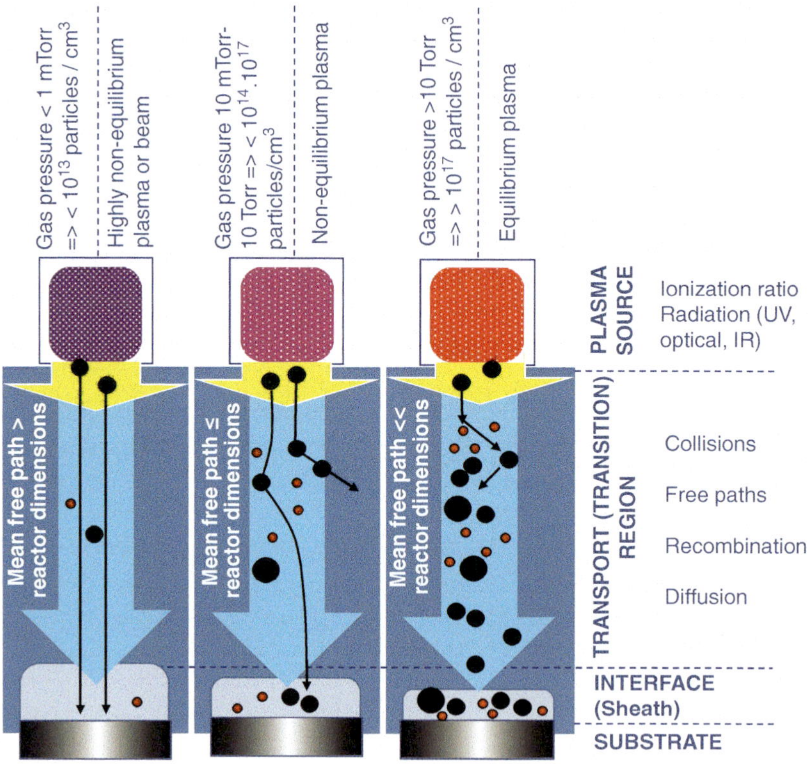


**Figure 3.4** Effects of the gas pressure on the particle transport in the plasma processing.

higher pressures, the probability of the particle collisions is higher, with all consequent changes of the density and composition of incoming particles to the substrate or, more correctly, to the surrounding sheath region with the heterogeneous reactions. As explained in Chapter 2, the character of the plasma changes from a low pressure non-equilibrium plasma to an equilibrium one.

## 3.4 Afterglow and Decaying Plasma Processing

In the plasma processing, the substrates are often positioned outside the plasma. There is often certain transition path over a transporting region between the area of plasma generation and the substrate surface, as was shown in Figure 3.4. The active particles generated in the plasma always have a certain life-time of activity (the relaxation time) as was shown in Figure 3.2. This lifetime depends on several parameters. Besides collisions (Figure 3.4), the chemical origin of given particle, its excitation level, ionization potential, etc. are important. The particles are capable to run only along a limited distance or during limited period of time. The plasma can generate high energy radiation, e.g. in UV range, which has limited reach. All such factors are important for the selection of distance between the substrate and the plasma.

Specific regions outside the main plasma generation are called *afterglows*. Two main types of the afterglows can be recognized: *time-resolved afterglows* and *space-resolved afterglows*. In these afterglows, we can recognize *decaying plasma*, *near afterglow*, and *far afterglow*. Schematic description of these parts is shown in Figures 3.5 and 3.6. Active particles generated in the plasma lose their activity during time or along the distance from the plasma. A typical example of the time-resolved afterglow is the pulsed plasma, where each pulse generates a certain density of active particles, and they decay after termination of each pulse. The steady-state density then depends on the duty-cycle, which is the ratio of durations of the pulse and the pause to the next pulse, see Figure 3.6.

An important factor in pulsed plasmas is the shape of the pulse, particularly the pulse rise-time. If the pulses are steep with a short rise-time, then at the beginning of each pulse, the plasma builds itself up or returns to the full parameter state after partial decay. At this stage, the ionization is not fully developed, and the voltages applied from the generators have not decreased enough. This state can generate a group of electrons accelerated to energies up to the total amplitude of the applied voltage. Such electrons can repeatedly overheat and damage the processed surface in the pulsed plasma, see Ref. [3.14]. Some consequences will be introduced in Section 4.1.2, Chapter 4. Therefore, the pulse geometry must be designed to avoid such negative effects.

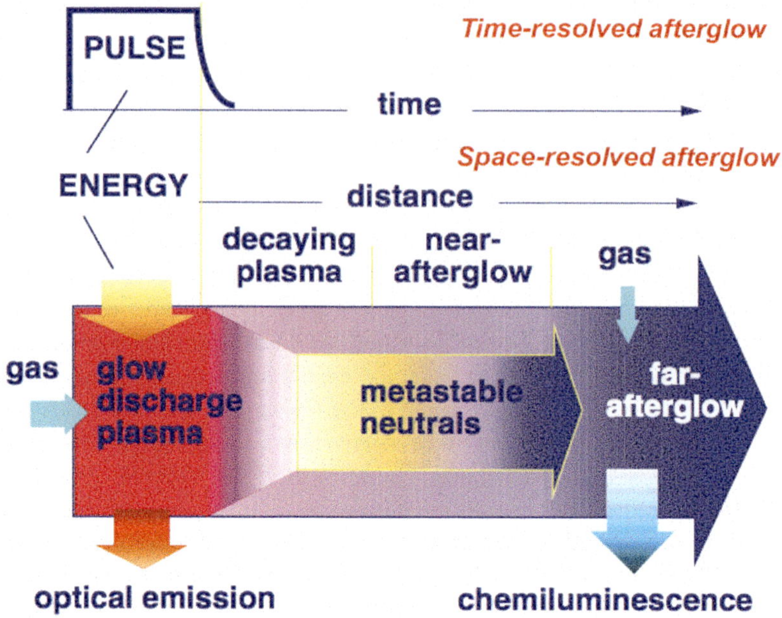


**Figure 3.5** Time-resolved and space-resolved afterglows. Certain gases can be excited by long-lived metastable particles from plasma. Recombination processes can produce visible light (chemiluminescence), see also Ref. [3.13].

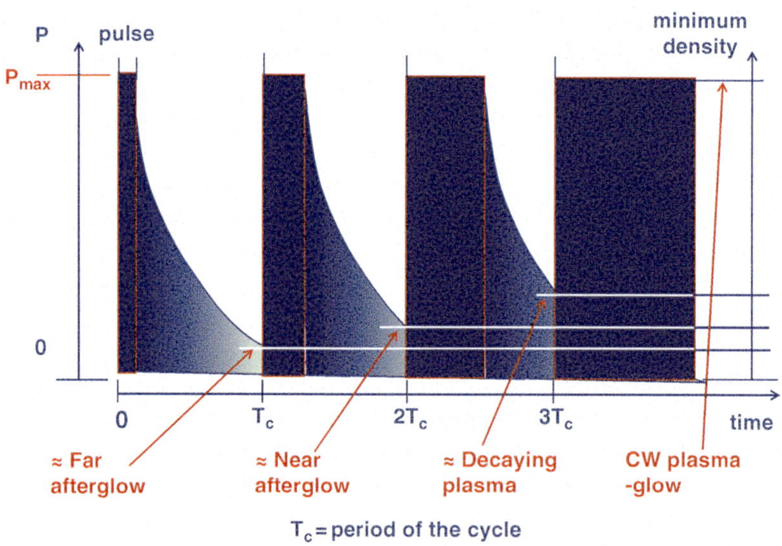


**Figure 3.6** The effect of the duty-cycle in a pulsed plasma on the minimum steady-state density of the plasma-generated particles.

In the space-resolved afterglows, the main importance has the position of substrates with respect to the plasma generation area (the plasma source). Density of the most energetic particles decreases with the distance from the plasma. The rate of this downfall (decrease) in a stationary gas increases with the increasing gas pressure, mainly due to increasing collision frequency. Of course, with cases of the forced flows or other gradients, this decrease can be partly compensated, which is valid particularly for the long-living metastable excited particles. The metastable particles are the most important ones in the region of far afterglow. An example of such particles is excited atomic nitrogen, which can be transported over relatively long distances, even up to several tens of centimeters, see, e.g. Refs. [3.15, 3.16]. An application of this nitrogen property in the afterglow deposition of silicon nitride films will be described in Section 4.2, Chapter 4.

## References

[3.1] M.A. Lieberman and A.J. Lichtenberg: "Principles of plasma discharges and materials processing", 2nd Edition, John Wiley & Sons, Inc, New Jersey, 2005.

[3.2] Y.A. Lebedev: "Microwave discharges in liquid dielectrics"; Plasma Phys. Rep. **43** (6) (2017) 685–695.

[3.3] J.F. O'Hanlon: "A user guide to vacuum technology", 3rd Edition, John Wiley & Sons, Inc, New Jersey, 2003.

[3.4] L. Bardos, R. Dragila, G. Loncar, and J. Musil: "Creation of thin oxide films in a microwave oxygen plasma", Rozpravy CSAV (in English), vol. **93**, no. 3, Academia, Prague, Czechoslovakia, 1983.

[3.5] S.M. Rossnagel, J.J. Cuomo, and W.D. Westwood, eds.: "Handbook of plasma processing technology", Noyes Publications, Park Ridge, New Jersey, 1990, in: Chapter 11, by J. Asmussen, p.305.

[3.6] B.M. Annaratone, V.P.T. Ku, and J.E. Allen; XXI. Int. Conference on Phenomena in Ionized Gases – ICPIG-21, September 19–24, 1993, Ruhr-Univ. Bochum, Germany, Proceedings ed.by G. Ecker, U. Arendt, J. Bösler, Proc. vol. **I**, pp. 29–30.

[3.7] R. Itatani: "Physics of reactive plasma for material tailoring"; Plasmas Ions **1** (1998) 37–44.

[3.8] J.W. Coburn: "Plasma etching and reactive ion etching", AVS Monograph Series, ed.by N.R. Whetten, Library no. 82-73342, Book nr. 0-88318-406-0, Publ.by the American Inst.Phys., Inc., New York, 1982.

[3.9] D.M. Manos and D.L. Flamm: "Plasma etching: An introduction (Plasma – Materials Interactions)", 1st Academic Press, Inc., Cambridge, Massachusetts 1989

[3.10] F.F. Chen: "Introduction to plasma physics", Plenum Press, New York, 1974.

[3.11] B. Chapman: "Glow discharge processes", John Wiley & Sons, New York, 1980, San Diego, 1989, Chap.2 – written by D.L. Flamm, Chap.3 – written by S.A. Cohen.

[3.12] A.S. Penfold and J.A. Thornton: U.S. Patent No. 4,116,794, priority 1978.

[3.13] L. Bardos: "Afterglow and decaying plasma CVD systems"; Vacuum **38** (8–10) (1988) 637–642.

[3.14] L. Bardos, R. Dragila, G. Loncar, and J. Musil: "Creation of thin oxide films in a microwave oxygen plasma"; ACADEMIA **93** (3) (Prague 1983).

[3.15] L. Bardos, J. Musil, and M. Lubanski: "Chemiluminescence of the silane – active nitrogen reactions during PECVD of the silicon nitride films"; Czech. J. Phys. B **34** (1984) 1242–1245.

[3.16] L. Bardos and J. Musil: "Axial decaying of the microwave ECR oxygen plasma"; J. Phys. D: Appl. Phys. **21** (1988) 1459–1461.

# 4

# Microwave Plasma Systems for Plasma Processing at Reduced Pressures

According to Figure 2.10 and Section 2.5.3 (Chapter 2), *the microwave discharges can be characterized as high density and low ion energy plasmas capable of forming reactive radicals and plasma-chemical processes.* The high electron densities microwave plasmas, particularly in the electron cyclotron resonance (ECR) and other magnetoactive plasmas, enable efficient control of particle energies and tuning these for selected reactions in the plasma processing. In this chapter, several microwave systems tested in the plasma processing will be described. Most of those were laboratory systems, but they can serve as an inspiration for larger systems. The microwave systems for generation of gas plasmas in technologies are based on similar principles and use similar components. This chapter deals only with the reduced- and low-pressure plasma systems. The atmospheric and high-pressure systems will be introduced in Chapter 5.

## 4.1 Waveguide-Generated Isotropic and Magnetoactive Microwave Plasmas

Pioneering experiments with the microwave plasma in waveguide structures were dated by the 1960s of the twentieth century. After the Nobel Prize in Physics was awarded for research in semiconductors and discovery of the transistor in 1956, the first works were devoted to the oxidation of semiconductors in an oxygen microwave plasma to create dielectric oxides at temperatures below 500°C, see Refs. [4.1–4.3]. The discharge system was arranged with a simple discharge tube about 2 cm in diameter passing through a tapered part of the rectangular waveguide (see Figure 2.11c, Chapter 2). With selected parameters (gas pressure, position of the tube), this arrangement can work as a resonator with low reflected power. Several waveguide-based resonator cavities (see Section 1.3.5, Chapter 1) for generation of discharges were described in 1965, see Ref. [4.4]. Some of these cavities were capable of generating an atmospheric microwave plasma. More complicated microwave structures were

*Microwave Plasma Sources and Methods in Processing Technology*, First Edition.
Ladislav Bárdoš and Hana Baránková.

tested with plasmas in the magnetic field, see Refs. [4.5–4.7]. Some of those will be described in Section 4.1.2. Different microwave plasma systems with details and rich lists of references can be found, see Refs. [4.8, 4.9]. The plasma generated without any auxiliary magnetic field can be considered as an *isotropic plasma* because no specific force affects the plasma parameters. This is contrary to an *anisotropic plasma* where the anisotropy is given by an auxiliary magnetic field. Such plasma can be characterized as a *magnetized or magnetoactive plasma*.

As explained in Chapters 2 and 3, the microwave plasmas often use auxiliary magnetic fields. The magnetic field of induction $B_{ce} = m_e\omega/e$, where $m_e$ and e are mass and charge of the electron, respectively, and $\omega = 2\pi f$ is the frequency (called angular frequency) of the microwave generating power, allows resonant absorption of the microwave power (ECR) and much higher plasma density than the critical (cut-off) density $n_{crit}$ defined in Chapter 2 in Eq. (2.10). For the microwave frequency f = 2.4 GHz, the $n_{crit} \approx 7.1\ 10^{10}\ cm^{-3}$ (for f = 2.35 GHz $n_{crit} \approx 6.8\ 10^{10}\ cm^{-3}$). In the ECR and higher induction magnetized microwave plasmas, the plasma density can reach an order of $10^{13}\ cm^{-3}$ at several mTorr of the gas pressure. This classifies such plasmas as *high-density plasmas*. The presence of magnetic field means a presence of a driving force for the charged particles in the plasma, defined by the Lorentz force vector, see Eq. (3.11) in Chapter 3. In Chapter 3, gradients of the non-uniform magnetic inductions were shown to represent extra driving forces for charged particles, see Eq. (3.14) in Section 3.3.2, Chapter 3. Because of Coulomb forces acting between electrons and ions, the driving of electrons or ions means driving the whole plasma in the same direction by an effect called the *ambipolar diffusion*. This explains why the magnetic field must be designed and positioned carefully. Several microwave magnetoactive plasma systems used in oxidation experiments, see Refs. [4.10–4.16], are schematically illustrated in Figure 4.1.

Large number of reported experiments with the microwave plasma in oxygen for oxidation of solid substrates, particularly silicon, revealed specific features of the microwave plasmas in their generation and their properties. Therefore, the microwave oxidation and anodization (positively biased substrate) could serve as a good introduction to the microwave plasma processing. The microwave plasma in oxygen and its mixtures has important applications in modern technologies, particularly in semiconductor production, see Refs. [4.17, 4.18]. The microwave oxygen plasma at 0.7–1.5 kW and 6 kPa (45 Torr) has been used for oxidation of 4H-SiC substrates for producing metal-oxide semiconductor field effect transistors (MOSFETs), see Ref. [4.19]. Another example is photoresist stripping by the microwave plasma in $O_2$ mixed with $CF_4$, or $N_2H_2$, see Ref. [4.20]. An interesting novel application is oxidation of gallium nitride (GaN) substrates, see Ref. [4.21]. Gallium nitride-based materials are of growing interest due to their potential for high-temperature,

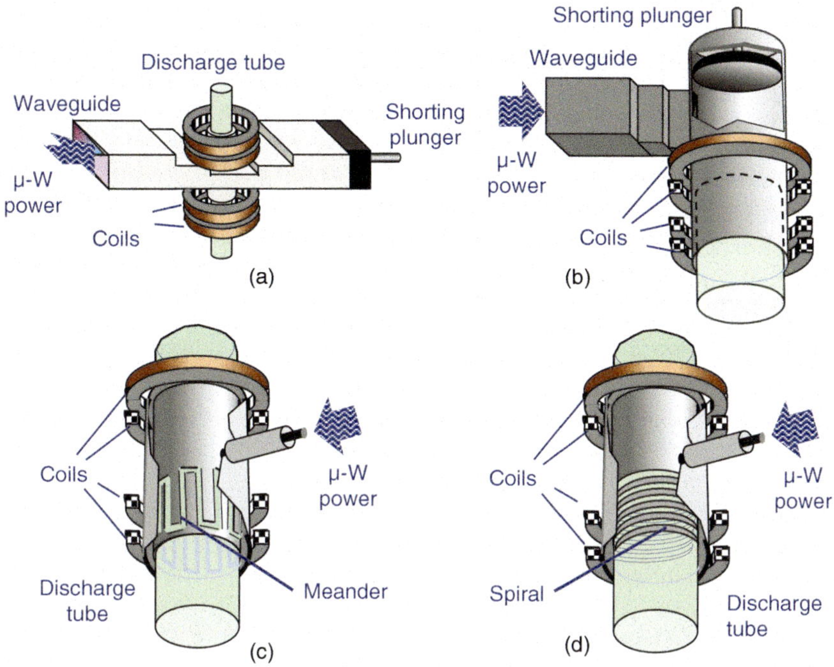


**Figure 4.1** Microwave magnetoactive plasma systems. (a) Rectangular waveguide system, (b) Circular waveguide terminated by a discharge tube, (c) Lisitano slow-wave structure with meander radiator, (d) spiral type radiator slow-wave structure.

high-power, and high-frequency electronic devices. The GaN devices could replace silicon-based semiconductor devices. Here the microwave oxygen plasma is a potential candidate for replacement of thermal and chemical oxidation methods of GaN substrates, see Ref. [4.22]. Like in the oxidation of silicon, the formation of native insulating oxides $Ga_2O_3$ on GaN at low oxidation temperatures (≤300°C) can become important in fabrication of metal-oxide semiconductor (MOS) structures with low interface state density and low leakage currents. In this chapter, the plasma oxidation and anodization (illustrated in Figure 3.3, Chapter 3) will be described in more details for silicon substrates, based on the authors' experimental experience.

### 4.1.1 Waveguide-Generated Isotropic Microwave Oxygen Plasma for Silicon Oxidation

The plasma in molecular gases is more complicated than the plasmas in atomic gases because of the additional power absorbing channels (e.g. in the molecular rotations and vibrations) and the formation of the excited and ionized radicals. One of the most complex molecular gases is oxygen because of its electronegativity with forming of atomic and molecular negative ions ($O^-$ and $O_2^-$). An important role of the negative oxygen ions has been proved by the oxidation experiments in the microwave oxygen plasma and the oxidation rate enhancements by applying positive bias on the silicon substrate, Refs. [4.1, 4.2, 4.10]. A theoretical model based on the diffusion of negative oxygen ions into the Si substrate with simultaneous decreasing of the diffusion due to growing $SiO_2$ dielectric and a certain level of sputtering/evaporation of the oxide agreed with the experiments, see Figure 4.2. This figure shows the time decrease of the DC anodization current, see Ref. [4.11]. A simplified schema of the plasma anodic oxidation of Si was shown in Figure 3.3, Chapter 3.

The formation of negative oxygen ions by electron attachment requires rather low electron energy (about 4–6 eV, see Table III in Ref. [4.23]) and the attachment is usually connected with the dissociation of the oxygen molecule according to the following reaction:

$$O_2 + e \rightarrow O^- + O \tag{4.1}$$

The maximum formation of the negative ions $O^-$ is around 5 eV, see Ref. [4.24]. These ions are not visible in the optical emission spectra from the microwave plasma, and their density can be measured only indirectly, e.g. from dissociation reactions between $O^-$ and $O^+$ ions in the afterglow regions, see Ref. [4.25].

Typical emission spectrum detected from the microwave oxygen plasma is shown in Figure 4.3. The detected spectra confirm molecular excitation and ionized oxygen molecules in a strong First Negative System of $O_2^+$ ions. The intensity of emission is enhanced by the negative bias of −500 V on the

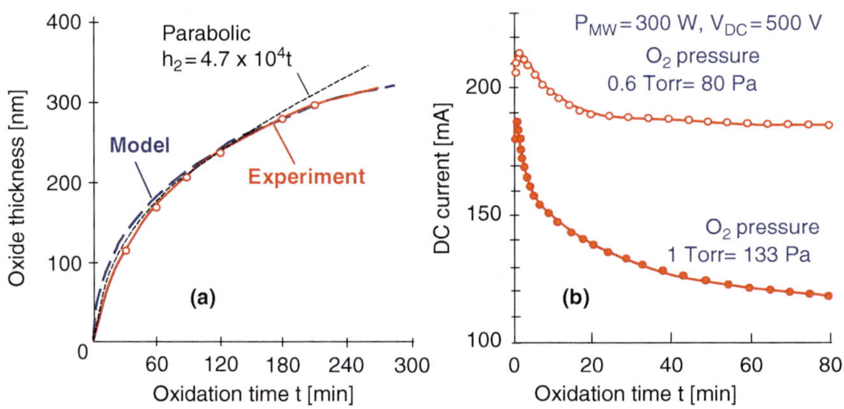


**Figure 4.2** Comparison of the theoretical model of the plasma oxidation with the experiment (a). Decreasing DC anodization current into Si substrate during oxidation time (b).

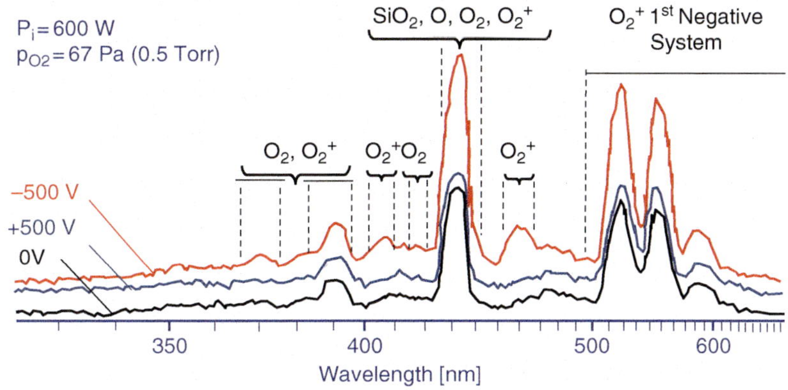


**Figure 4.3** The optical emission spectra of the microwave-generated oxygen plasma at 0.5 Torr and 600 W of the power. The spectra were acquired for three DC voltages between inserted electrodes. The zero levels of the spectra are shifted to avoid coincidences.

electrode positioned in close vicinity of the detected area, see the test system in Figure 4.4. The counter electrode was grounded. The bias on Si cathode can form secondary electrons and enhance density of the plasma. The cathode bombardment by positive ions probably caused Si sputtering and forming of SiO bonds. The most intense band head in the spectrum coincides with the strongest line of the oxygen atom at 436.8 nm.

The test system with the tapered waveguide section for increasing the microwave electric field in a quartz discharge tube is shown in Figure 4.4a. The measuring line with the probe allowed detection of the standing wave ratio

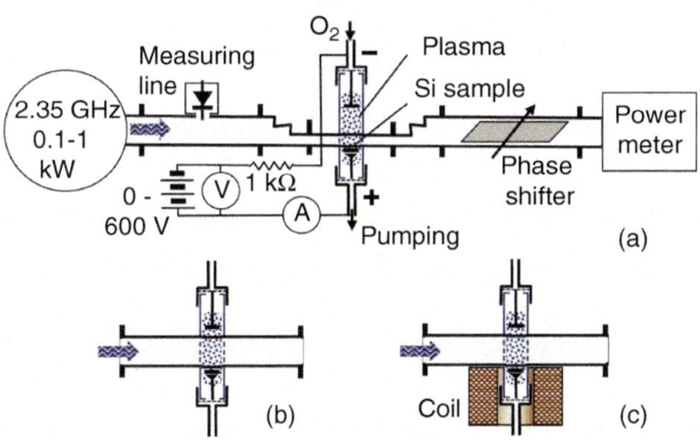


**Figure 4.4** Schematic of the waveguide system for microwave generation of oxygen plasma in 40 mm outer diameter tube (a). A discharge section in the untapered waveguide (b). An arrangement with magnetic coil for study of motions of charged particles (c).

(SWR), see Section 1.3.3, Chapter 1. The waveguide power line was terminated by the water calorimeter (power meter). The electrodes installed in the system were connected to a DC source with adjustable voltage for the anodization regimes. An arrangement without tapered waveguide section, see Figure 4.4b. This version was arranged with a magnetic coil, see Figure 4.4c, to study the magnetic effect on the charged particles. The photograph in Figure 4.5 shows the arrangement illustrated in Figure 4.4a. Figure 4.5 shows the discharge tube with 25 mm in diameter supported by insulating rods on both sides of the tapered part of the waveguide. The tapered waveguide in the power line was installed between two quarter-wave transformers. The principles of such transformers were described in Section 1.4.6, Chapter 1, and illustrated schematically in Figure 1.19b, Chapter 1.

The microwave system shown in Figure 4.4 usually generated an axially asymmetric plasma, clearly visible at low powers. Because the power was delivered from one side, it was almost impossible to tune fully symmetric plasma in 40 mm diameter tubes, and the oxide area on the Si sample had always higher thickness at the power side, see Figure 4.6.

The search for the most important particles in the growth of Si oxides revealed the possibility of affecting oxide areas by an auxiliary magnetic field.

**Figure 4.5** Detail of the experimental arrangement with 25 mm diameter discharge tube.

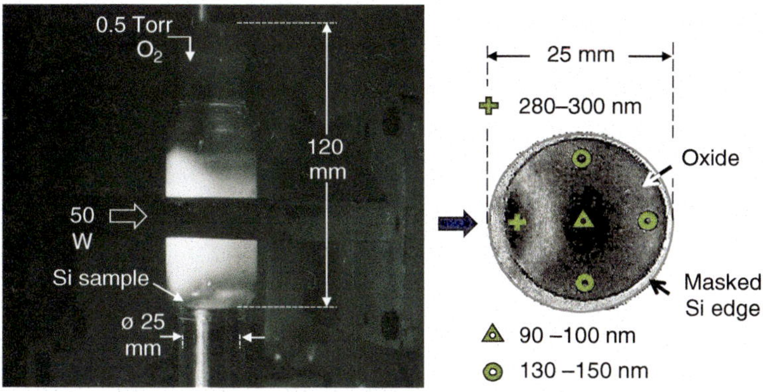


**Figure 4.6** The non-uniform oxygen microwave plasma led to a non-uniform distribution of the oxidized surface on 25 mm diameter Si substrates in 40 mm diameter discharge tube.

If the negative oxygen ions have an important role in the silicon oxidation, such ions should be produced by an electron attachment at the surface of the substrate or in the adjacent sheath. This hypothesis was experimentally confirmed by using a magnetic coil around the sample, as shown in the photograph in Figure 4.7.

Figure 4.8 shows the arrangement and the geometry of the oxide spots obtained at different magnetic fields. The oxide spot was extended to one side

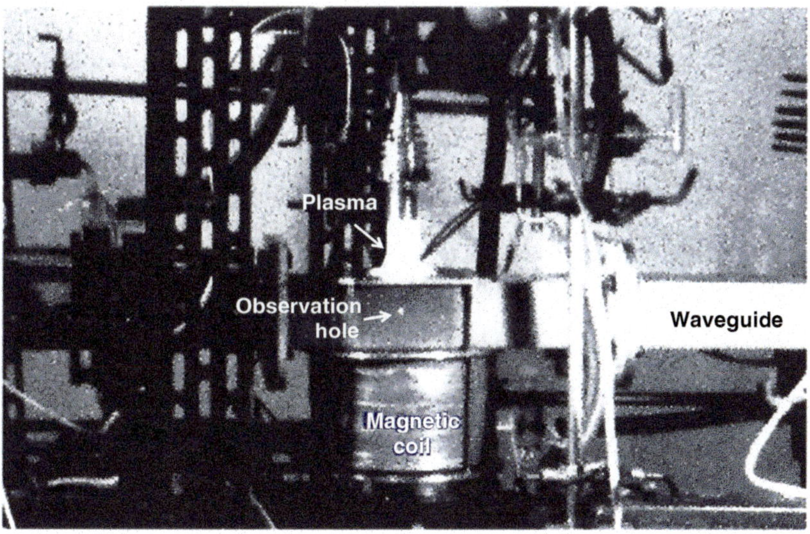


**Figure 4.7** Experimental arrangement with the magnetic coil surrounding the bottom part of the discharge chamber with the silicon sample.

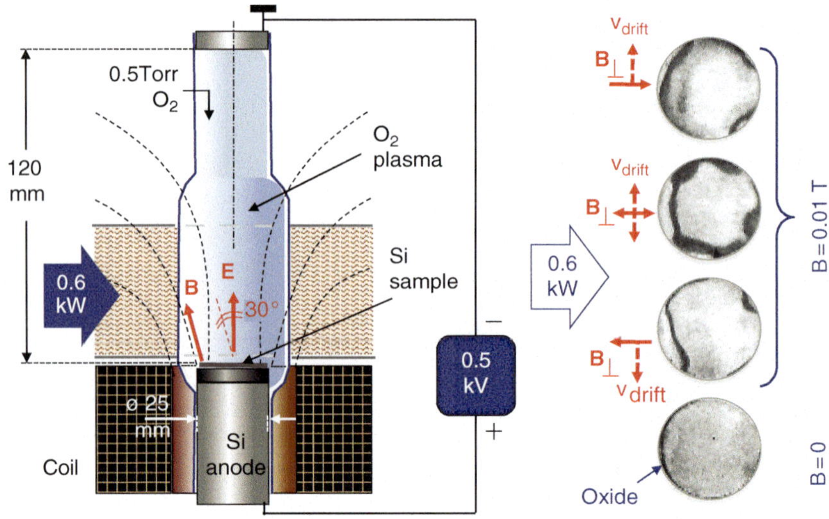


**Figure 4.8** An illustration of the effect of auxiliary magnetic field on the geometry of oxidized silicon surface. The field orientation was changed by the polarity of DC current in the coil.

on the sample surface, depending on the orientation of the induction vector changed by the polarity of the DC current in the coil. When the AC current was used with the same amplitude as the DC current in the magnetic coil, the oxide spot was extended symmetrically to both sides. The magnetic induction vector **B** at the edge of the Si slice had an angle of about $\alpha \approx 30°$ from the axis, i.e. from the vector **E** generated by the DC voltage applied between the sample and a grounded counter electrode. The induction value corresponding to the magnetic component perpendicular to the electric field was calculated as $B_L = B \sin \alpha$ from $B \approx 0.01$ Tesla (= 100 Gauss). The intensity of the electric field between the sample and a counter electrode was $E \approx 500$ V/12 cm $\approx 4.17$ kV/m.

From the calculated drift velocity and Larmor radii with respective masses of the electrons and $O^-$ and $O_2^-$ ions, there were only electrons which could fit with the size and the shift of the oxide spot, see Ref. [4.16]. This experiment proved the main role of electrons in the oxidation process and their responsibility for the generation of oxygen negative ions on the surface of the silicon sample. Therefore, electron density and its uniformity over the sample surface are the two most important parameters defining the oxidation rate and the uniformity of the resulting oxide, and the formation of negative oxygen ions requires proper energy of electrons (roughly between 2 and 10 eV). Experiments in high-density pulsed magnetoactive microwave plasmas confirmed that the electron energy is another limiting factor in the rate of the oxidation and the properties of the oxide layer.

### 4.1.2 ECR and Higher Induction Magnetized Plasma Systems for Silicon Oxidation

Diameters of the discharge tubes passing rectangular waveguides (Figure 4.1a) are naturally limited by the cross-dimension of the waveguide. The maximum acceptable diameter was found to be up to about 40 mm. For larger diameter plasmas it is necessary to use different arrangements of the microwave systems. The discharge tube having 84 mm inner diameter has been used for generation of an oxygen magnetoactive plasma.

The system shown in Figure 4.1b was tested with the microwave magnetron generator giving maximum power up to 3 kW at $f = 2.35$ GHz ($\omega = 14.76$ GHz). This magnetron allowed pulsed regimes with the repetition frequency of 50 Hz and with the pulsing duty-cycle controlled by the tunable pulse length. Typical pulse length used in the experiments was 3 ms. Application of the magnetic field at the ECR, i.e. for $\omega = \omega_{ce}$ and/or $B = B_{ce}$ ($B_{ce} = \omega\ m_e/e \approx 8.4 \times 10^{-2}$ Tesla = 840 Gauss), and the ECR field (up to $B/Bce \leq 3$) allowed using large-diameter discharge tubes and, consequently, the larger diameter silicon substrates. Moreover, it was possible to reduce the oxygen pressure substantially (to $\leq 0.1$ mTorr), which expanded the volume of the plasma (longer mean-free paths).

The power functions of the magnetoactive plasma density were measured by the 8 mm microwave interferometer, and the corresponding heating of the silicon sample was measured by the thermocouple integrated in the substrate holder. The results revealed favorable parameters of the pulsed power regime over the continual wave (CW) regime, see Figure 4.9, which shows that at the same incident power, the pulse regime generated about three times higher plasma density and about three times lower heating of the Si substrate. The substrate heating can be caused by the bombardment of particles (depending on the plasma potential and particle energies) and by the microwave power absorbed in the silicon substrate (see Table 1, Chapter 1, Section 1.4.7).

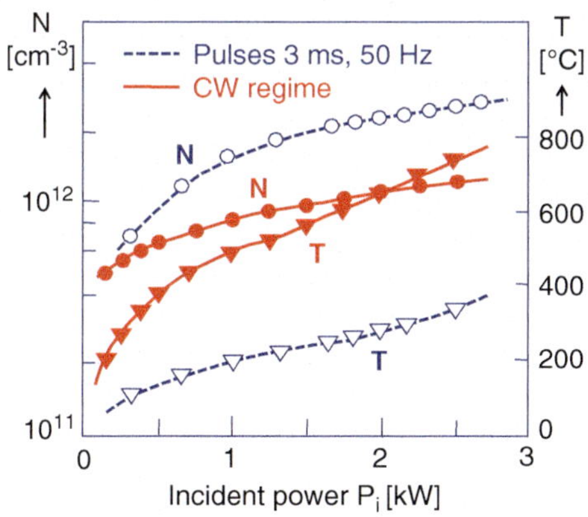


**Figure 4.9** Comparison of the power dependencies of the plasma density and heating of the Si sample in the pulsed regime and in the continuous power regime. The magnetic field $B = 2.2\ B_{CE}$, oxygen pressure $p_{O2} = 0.27$ Pa (2 mTorr). No DC bias was applied on the Si sample.

Experiments in several gases (i.e. $H_2$, He, Ar, $O_2$) showed that the generation of the microwave magnetoactive plasma strongly depends on the value of the magnetic induction. An easy start of the plasma by the microwave power around 1 kW was possible only at the ECR condition, i.e. in the magnetic field of $B = B_{ce}$. As proved in the hydrogen plasma at 0.5 mTorr, for the inductions below and above the $B_{ce}$ value the plasma needed certain pre-ionization, see Ref. [4.26]. To prove this, the system was arranged by two slow-wave Lisitano helical structures (see Figure 4.1d) working simultaneously on the same discharge tube (inner diameter of 80 mm). One structure was powered by 2.35 GHz at 2 kW and the other one by 1.25 GHz at 5 kW. For the lower frequency f = 1.25 GHz, the corresponding ECR induction value is $B_{ECR} = 2\pi\ 1.24 \times 10^9 \times 9.11 \times 10^{-31} \times 10^{19}/1.6 = 3.58\ 10^{-2}$ Tesla = 358 Gauss. The magnetic inductions were tested in the interval up to $\omega_{ce}/\omega \approx 14$ for both frequencies.

In high magnetic fields, exceeding the ECR value, the microwave power was absorbed efficiently, and the plasma density reached almost 80 times the critical density $N \approx 80\ n_{crit}$. (The critical or cut-off density $n_{crit}$ was described in Chapter 2, Eq. (2.10)). For 2.35 GHz, the critical density is $n_{crit} \approx 6.8\ 10^{10}\ cm^{-3}$, and the obtained plasma density was two orders of magnitude higher $N \approx 5.4\ 10^{12}\ cm^{-3}$.

The helical or meander type of the microwave slow-wave structures (shown in Chapter 4.1, in Figure 4.1c, d, Section 4.1) for the generation of dense magnetoactive plasmas were invented by Giuseppe Lisitano and further developed by his co-authors, see Refs. [4.5, 4.6]. The principles of operation of such structures were described in Ref. [4.27]. The Lisitano structures were suitable for the high-density oxygen plasma for the oxidation of silicon. The photograph of the test system in operation is shown in Figure 4.10.

Systematic experiments were performed in oxygen at pressure of 1 Pa (7.5 mTorr), at 1.6 kW microwave power (f = 2.35 GHz), with the 3 ms pulses at 50 Hz repetition frequency. The axial magnetic field had the ECR induction value, i.e. $B = B_{ce}$ (or $\omega_{ce}/\omega = 1$) and the plasma density was detected by the 8 mm microwave interferometer. These experiments tested different diameters of the microwave structures with corresponding diameters of the discharge tubes (i.e. 23, 33, 45 and 75 mm) passing through their axes. It was found the plasma density N depended almost inversely on the diameter of the plasma column, see also Ref. [4.12]. These data confirmed the plasma density was proportional to the surfaces of the plasma columns and not to their volumes. The corresponding graphical results are shown in Figure 4.11.

It was found in the tube diameter of 33 mm that the plasma density grows with the incident microwave power and with the magnetic field, see Figure 4.12. The results shown in this figure were found at oxygen pressure of 1 Pa. The high magnetic fields evidently increased the power absorption efficiency

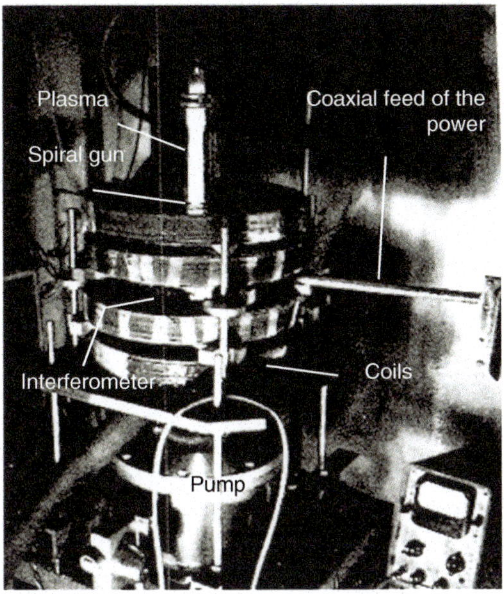


**Figure 4.10** The simple test system for generation of the microwave magnetoactive plasma by the spiral-type Lisitano slow-wave structure in the axial magnetic field.

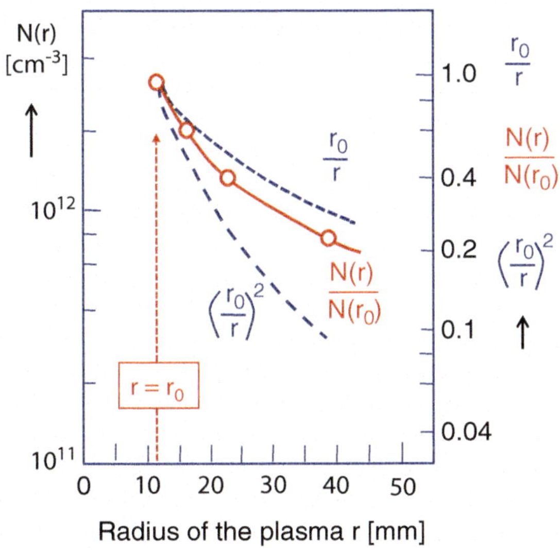


**Figure 4.11** The plasma density vs radius of the plasma in the discharge tubes with diameters of 23, 33, 45 and 75 mm installed in the spiral slow-wave structures with corresponding diameters. Incident power $P_i$ = 1.6 kW, 3 ms pulses at 50 Hz, $B = B_{ce}$, $p_{O2}$ = 1 Pa (7.5 mTorr).

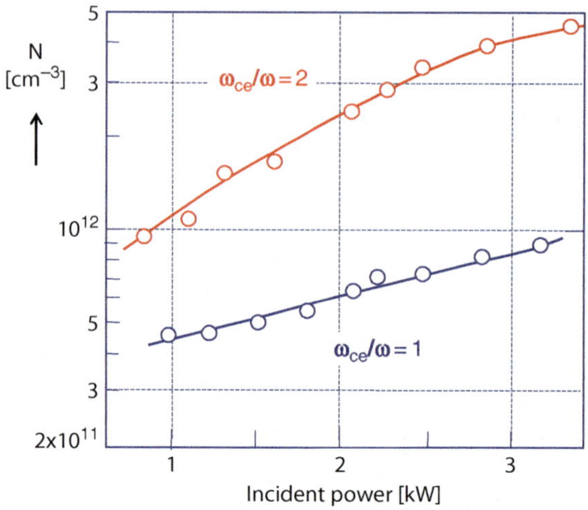


**Figure 4.12** The comparison of dependencies of the plasma density on the incident microwave power at the ECR magnetic field and at two times higher magnetic induction. The tube diameter was 33 mm, the oxygen pressure was $p_{O2}$ = 1 Pa. The pulsed power used 3 ms pulses at 50 Hz repetition frequency.

and allowed generation of dense plasmas at gas pressures ≥ 0.1 Pa (≈ 0.8 mTorr). This was particularly efficient in smaller diameter discharges (tubes).

As mentioned, an advance of efficient microwave absorption at high magnetic fields under stable pulsed power and with instant pre-ionization were successfully applied in the oxidation experiments. The sophisticated magnetoactive plasma system built for these experiments used two Lisitano guns (see Figure 4.1c, d) operated at two microwave frequencies, 2.35 GHz up to 3 kW and 1.25 GHz up to 5 kW, in both pulsed (3 ms, 50 Hz) and continual regimes. Schematic picture of this experimental system is shown in Figure 4.13.

The system in Figure 4.13 was tested at one power: 2.35 GHz. An important finding was that the plasma generation by CW power could operate at high magnetic fields without pre-ionization. At a sufficiently high microwave power, starting an ECR plasma (electron density $N \geq n_{crit}$) was possible when the magnetic field intensity was increased after the ECR plasma was on. In this case, the plasma generated at ECR served as the pre-ionization for the higher fields. Similar system to that in Figure 4.13 was described in Ref. [4.26]. The ECR region for 2.35 GHz ($B_{ECR}$ = 840 Gauss) is marked in Figure 4.13 by a hatched area. For the lower frequency f = 1.25 GHz, the corresponding ECR induction value is $B_{ECR}$ = 358 Gauss. The distribution of the magnetic induction in the axis of the discharge tube was controlled by the DC current in the coils. The coils were electrically connected in parallel.

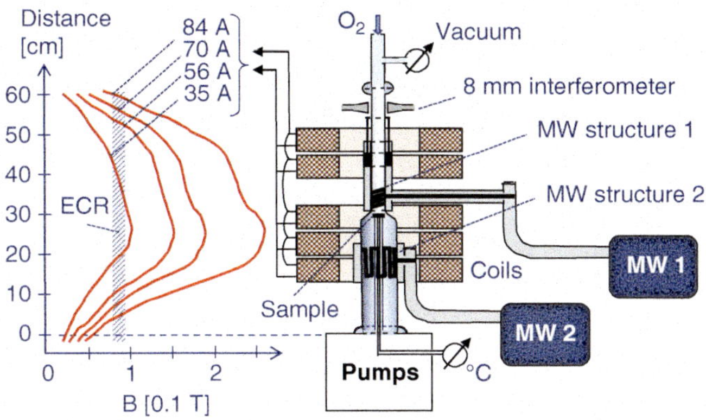


**Figure 4.13** Schematic of the experimental system for oxidation of silicon samples in a magnetoactive microwave plasma. Diameters of the quartz discharge tube were 33 mm (upper part) and 80 mm (lower part). Maximum induction in the axis of the tube increased to 4 $B_{CE.}$

In the power with CW regimes, the Si oxidation rates reached very high rates, as shown in Figure 4.14. At microwave powers P > 1.5 kW, the oxidation rate differed from the theoretical parabolic shape. This was caused by superpositions of the parabolic oxidation with the linear reactive deposition of $SiO_2$ from the partially evaporated edges of the Si sample.

The partial evaporation of the overheated sample edges was caused by fast electrons generated in the plasma by parametric instabilities and by the microwave power absorption in the sample. The evaporation of the Si sample edges was apparent on the series of the 25 mm diameter samples oxidized in the CW power, with example shown in Figure 4.15.

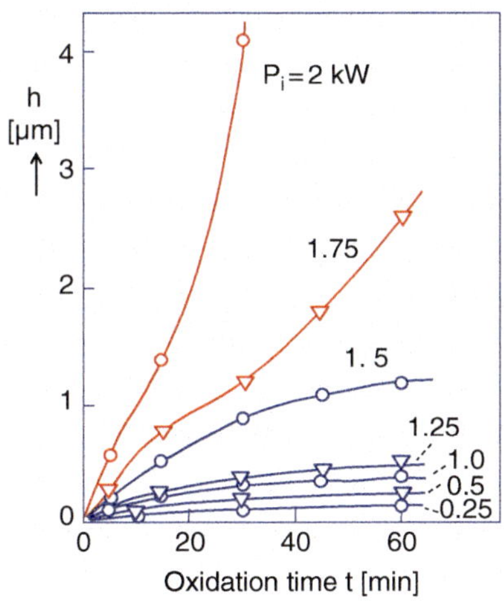


**Figure 4.14** The oxide thickness vs time on 25 mm diameter silicon samples in 2.45 GHz CW powered oxygen plasma at $B_{ce}/B$ = 2.2 and oxygen pressure $p_{O2}$ = 0.27 Pa (2 mTorr).

**Figure 4.15** Oxides on Si samples vs oxidation time at 2 kW power. Oxidation parameters are same as in Figure 4.14. One interference color ring corresponds roughly to 100 nm thick oxide.

At different plasma densities generated in the oxygen plasmas, different floating potentials $U_{fl}$ were measured. Floating potentials shown in Figure 4.16 were measured on an insulated (floating) probe with respect to the ground in the axis of the plasma. As introduced in Chapter 2.4, Eqs. (2.19 and 2.20) (Section 2.4, Chapter 2), *the floating potential ($U_{fl}$) of an electrode in the plasma represents the voltage with respect to the ground when the fluxes of electrons and ions are balanced and the resulting current to the electrode is zero*. Such potential should be detected on insulated (i.e. electrically floated) Si substrates. However, if the sample is placed on an electrically grounded holder, the measured potential will represent the potential of the plasma (the *plasma potential*) with respect to the grounded sample surface.

The plasma density in Figure 4.16 was measured by 8 mm microwave interferometer and represented an average density across the plasma column. The oxygen plasma was generated here by the pulsed microwave power up to 3 kW (f = 2.35 GHz, 3 ms, 50 Hz). The oxygen pressure was 0.4 Pa (3 mTorr) and the magnetic field was $B = 1.43\ B_{ce}$. An important coincidence was found between the floating potential and the ability of oxygen plasma to oxidize electrically grounded Si samples. As shown in Figure 4.16, the oxidation was observed

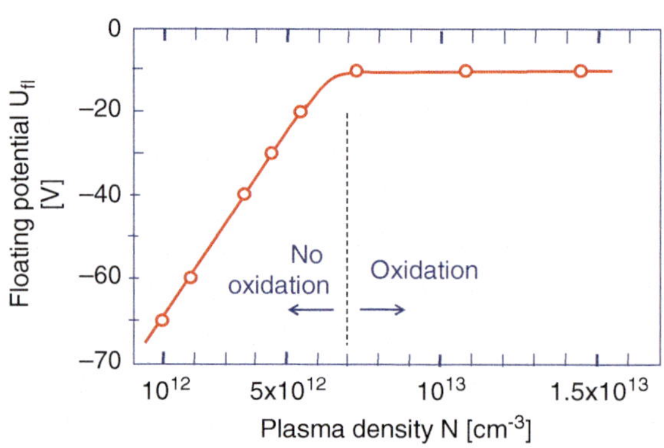


**Figure 4.16** The floating potential vs the plasma density. Parameters: $P \leqslant 3$ kW (3 ms, 50 Hz), $p_{O2} = 0.4$ Pa, $B = 1.43\ B_{ce}$.

only when the floating potential was not more negative than about – 10 V. The more negative potentials indicated a higher flux of electrons to the grounded surface and higher electron energy. The presence of low energy electrons forming negative oxygen ions for oxidation could be considerably reduced here. Moreover, the incidence of energetic electrons could even overheat the sample and destroy the oxide (such effect will be discussed again later).

Figure 4.17 shows that the floating potential strongly depended on the magnetic induction and could reach even positive values favorable for the oxidation.

Figure 4.17 shows that the plasma was unstable between induction $B_{CE}$ (ECR induction) and 2 $B_{CE}$. In this region, the oxide growth was stopped completely, presumably due to high electron energies. Negative effects caused by fast electrons were found in all oxides grown in the pulsed plasma, see Ref. [4.13]. Contrary to the oxidations in the CW plasma, the pulsed plasma resulted in partly damaged oxides having low values of the oxide dielectric strength and great density of the oxide defect charges. This was probably caused by the formation of electrons with energies exceeding 100 eV. Such electrons can be generated particularly at oxygen pressures under 0.67 Pa (5 mTorr) in the magnetic fields B > 2 $B_{CE}$. Previous studies of the pulsed ECR plasmas in helium and hydrogen at 0.18 Pa and 0.08 Pa (1.35 and 0.6 mTorr) with 0.19 ms 50 Hz 2 kW power pulses revealed, that at the beginning of each pulse with the steep rise-time (shorter than about 5 μs), the plasma only starts to build up and

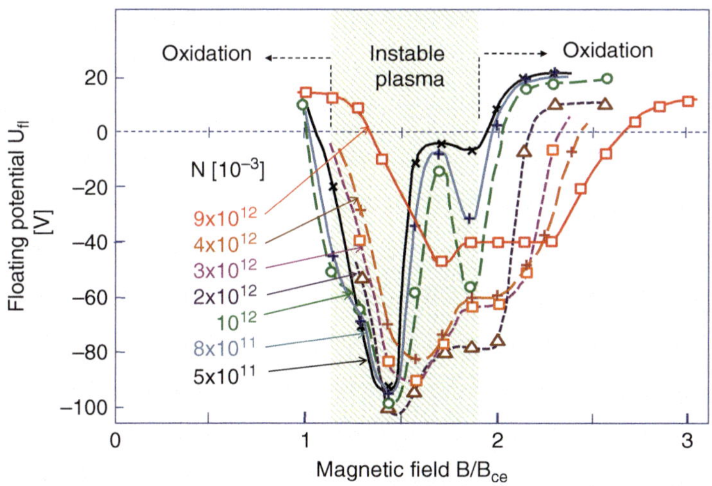


**Figure 4.17** Dependence of the floating potential in the oxygen plasma on the magnetic induction. The plasma density was controlled by the level of the pulsed microwave power up to P = 3 kW (3 ms, 50 Hz). Oxygen pressure was $p_{O2}$ = 0.4 Pa.

the electrons can easily acquire energies of the order of 1 keV, see Refs. [4.27, 4.28]. Similar observations were recently confirmed in Ref. [4.29], where the author wrote:

"*EEDF is greatly enriched by the fast electrons in the initial phase of the pulsed discharge in helium, and relaxes to a stationary value over some time, due to the processes of the inelastic interactions of the electrons with the gas atoms*".

Avoiding the generation of fast electrons is possible by proper "shaping" of the power pulses either with a slower or with a step-wise power increase. This idea needs more detailed experimental verifications.

The uniform thick oxides with properties of standard thermal oxides were obtained in the arrangement shown in Figure 4.1b. An oxide with a thickness uniformity of ≤ 5% formed on a 40 mm diameter Si slice in a discharge tube of 90 mm in diameter is shown in Figure 4.18.

The experimental arrangement used for the uniform oxidations is shown in Figure 4.19. The power in this arrangement was applied by the circular waveguide from the top of the discharge tube, see also Figure 4.1b. The sample position during oxidation was in a decaying plasma region or in a near afterglow. (Afterglows were explained in Section 3.4, Chapter 3.).

To find saturated ion currents and floating potentials typical in the decaying magnetoactive plasmas the sample holder was replaced by movable probes. This study is important also for other gases than oxygen, and the results were published in Refs. [4.30, 4.31]. The used probe had two forms of active surface: a small wire (Figure 4.19b) or a flat plate from different metals (Figure 4.19c). The probe potential was strongly negative ($V_{probe} = -100$ V), therefore, only positive ions ($O^+$ and $O_2^+$) contributed to the measured currents.

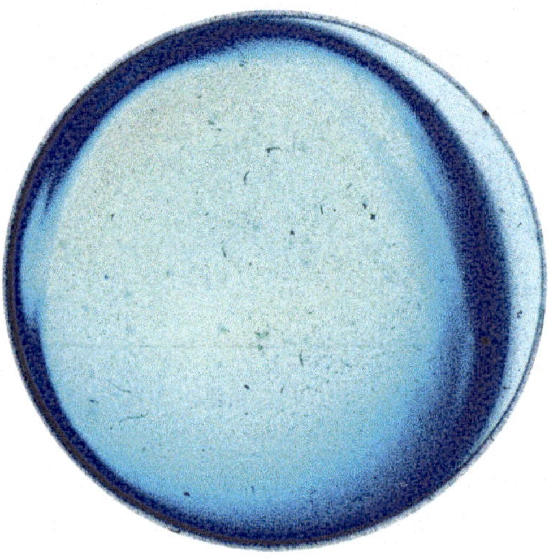

**Figure 4.18** A uniform oxide formed on 40 mm diameter Si substrate in the oxygen magnetoactive plasma.

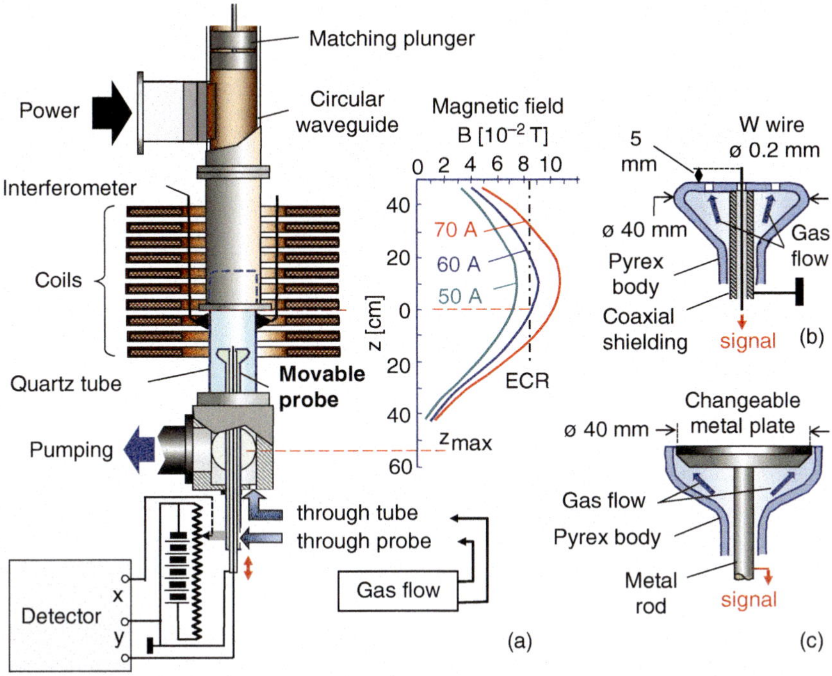


**Figure 4.19** Experimental arrangement for the oxidation of silicon and for test measurements of parameters of the magnetoactive plasma by movable probes. The distance z is measured from the termination of the circular waveguide (z = 0). The movable probe was arranged with the W wire sticking of the glass table (b) or with a solid plate made from different materials (c).

The saturated ion currents collected by the tungsten wire probe (shown in Figure 4.19b) along the axis of the plasma showed an axial decaying of the plasma density along the decreased magnetic induction from the ECR region down, see Figure 4.20. The plasma density decayed in a similar way as the gradient of the induction B.

An effective area of the W wire collecting the ion current at −100 V was $A_p = 0.2 \times 5 \times 10^{-6}\ m^2$ and the plasma density N can be estimated from the saturated ion current $I_s$ to the probe by the following equation, see Ref. [4.32]:

$$I_s = 0.6\ N\ A_p \left(e\ E_e / m_i\right)^{1/2} \tag{4.2}$$

Here, e is the electron charge, $E_e$ is the electron energy in eV, and $m_i$ is the ion mass. If we estimate the electron energy to $E_e = 10$ eV, simple calculations for the saturated ion current $I_s = 6\ 10^{-3}$ A (6 mA in Figure 4.20) in the ECR region at 1 kW power give for the positive molecular oxygen ions ($m_{O2}{}^+$ is 5.32 $10^{-26}$ kg) the plasma density of $N \approx 3.75\ 10^{12}\ cm^{-3}$. For the positive atomic ions of

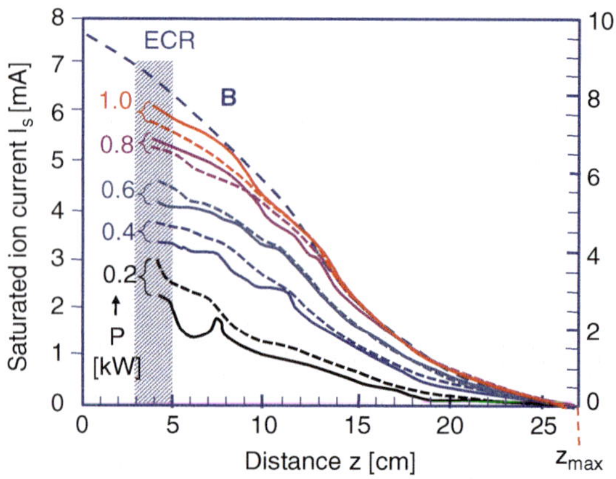


**Figure 4.20** Axial decaying of the saturated ion current (at $V_{probe} = -100$ V) for different microwave powers. Full lines are for the gas inflow from the bottom of the tube, and dashed lines are for the inflow through the probe. Oxygen pressure was $p_{O2} = 0.27$ Pa (2 mTorr).

oxygen ($m_O^+$ is 2.66 $10^{-26}$ kg), the density is 0.7 ($= 1/2^{-1/2}$) times less, i.e. $N \approx$ 2.63 $10^{12}$ $cm^{-3}$. If we estimate the electron energy as high as $E_e = 100$ eV, then these densities would be $N \approx 1.17$ $10^{12}$ $cm^{-3}$ for the molecular ions and $N \approx 8$ $10^{11}$ $cm^{-3}$ for the atomic oxygen ions, respectively. Because the high energy electrons are regularly observed in the ECR region, the energy estimate to 100 eV is quite realistic. What is of interest, the density measurements by the 8 mm microwave interferometer shown in Figure 4.9 for the CW power of 1 kW indicated an average plasma density of 8 $10^{11}$ $cm^{-3}$. Although this comparison is somewhat speculative, it indicates the main role of atomic ions. The presence of oxygen atoms available for ionization can be expected due to low dissociation energy of the $O_2$ molecules (only 5.15 eV), and the ionization potential of 13.5 eV for atomic ions $O^+$ is comparable with 12.2 eV for the molecular ions $O_2{}^+$.

In Chapter 3.3, in Eq. (3.14), it was stated that the gradient **B** represents a certain force, which moves charged particles from the higher to the lower field by an ambipolar diffusion. In an axially symmetric magnetic field, the radial loss of charged particles is limited, and the plasma density should follow the geometry of the applied magnetic induction. This effect can be observed mainly at lower gas pressures where the collision frequency is lower. The divergent plasma geometry in the decaying magnetic induction B is shown in Figure 4.21.

Figure 4.17 showed that, at oxygen pressure of 0.4 Pa, the floating potential depended on the magnetic induction. For the magnetic inductions below ECR ($B/B_{ce} < 1$), the floating potential was positive, which was found good for the growth of oxides on the Si substrates. The potential $U_{fl}$ was measured also at slightly lower oxygen pressure (0.27 Pa) by the electrically floating movable W

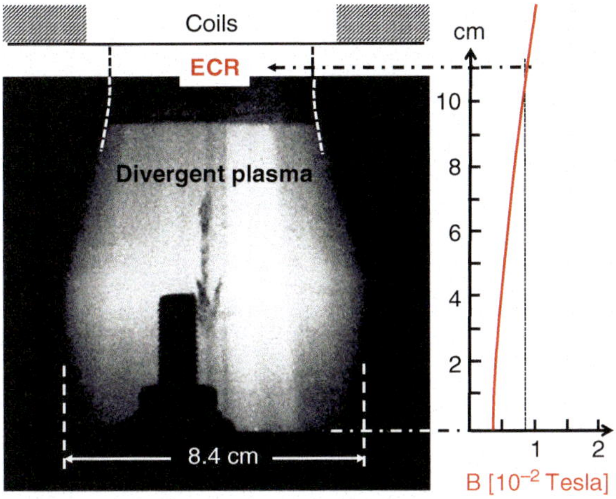


**Figure 4.21** A divergent geometry of the microwave magnetoactive plasma in the decaying magnetic induction. Incident microwave power was 1 kW, oxygen pressure was 0.27 Pa (2 mTorr). The quartz discharge chamber was cylinder with diameter of 84 mm.

probe (shown in Figure 4.19b) in the axis of the discharge tube for several microwave powers from 0.2 kW to 1 kW CW, see Figure 4.22.

The ECR region in the plasma was unstable. The floating potential had a tendency to fall to lower values, where higher electron fluxes required more ions to balance negative electrons. As expected, the changes in the floating potential were more pronounced at higher powers. Because the flux of charged

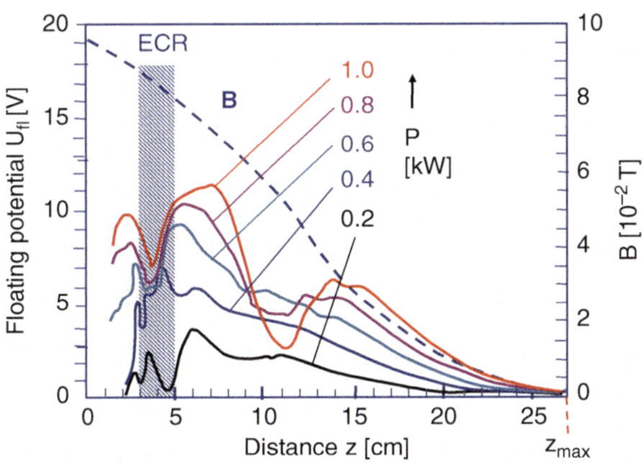


**Figure 4.22** Floating potential of the W probe along the decaying magnetic induction in the oxygen magnetoactive plasma generated at several powers. The oxygen gas was admitted through the bottom of the discharge chamber and its pressure was $p_{O2}$ = 0.27 Pa.

particles depends on their density and energy, the higher incident power can increase both these parameters and affect more the balancing conditions.

The floating potential strongly depended also on the oxygen pressure. Figure 4.23 shows results of the W probe measurements in oxygen at the microwave power of 0.5 kW. In these measurements, the oxygen was admitted from the bottom of the discharge reactor. The floating potential was more positive at lower pressures, i.e. at the lower collision frequencies. The values $U_{fl}$ were measured in the axis of the reactor, but the absorption of the microwave power in the plasma and, consequently, the plasma parameters depend also on the radial positions, see Ref. [4.33]. The radial non-uniformities of the plasma density and the floating potential were more pronounced in the ECR region and for $B/B_{CE} > 1$. Figure 4.23 shows how the $U_{fl}$ values depended on the gas pressure.

To involve larger radial plasma areas into the measurements the floating potential was measured by the flat probe (shown in Figure 4.19c). The results are in Figure 4.24. To interpret the results, for the higher secondary electron emission coefficient $\gamma$ (identical with the Townsend coefficient $\gamma$, see Section 2.2, Chapter 2), the more informative can be the floating potential $U_{fl}$, to limit the incoming ions creating the secondary electrons at the surface. As it can be seen in the measurements shown in Figure 4.24, the potential decreases in the following order of electrode materials: W, Cu, Al, Mo, Si and Fe. This order follows the maximum secondary electron emission coefficient $\gamma_{max}$ of the respective probe materials, with exceptions for Al and Fe. According to the Refs. [4.34, 4.35], the $\gamma_{max}$ values are the following: 1.4 (W), 1.3 (Cu), 1.3 (Fe), 1.25 (Mo), 1.1 (Si) and 1.0 (Al). The exceptions for Al and Fe can be explained by quick oxidation leading to different $\gamma_{max}$ on their oxidized surfaces. Indeed, the emission coefficient for $Al_2O_3$ depends on its thickness, and it can reach

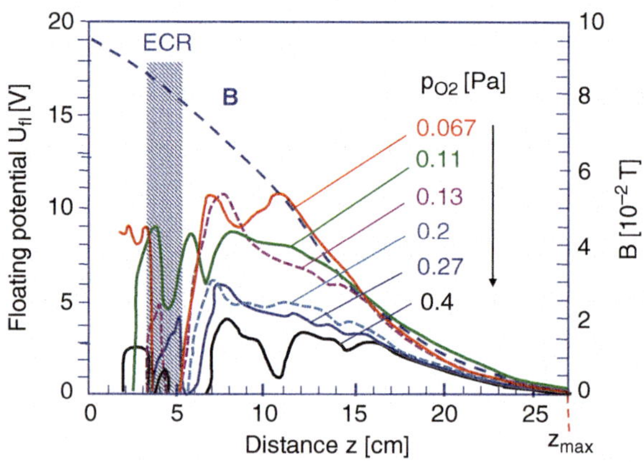


**Figure 4.23** Floating potential measured in the axis of the plasma at different oxygen pressures.

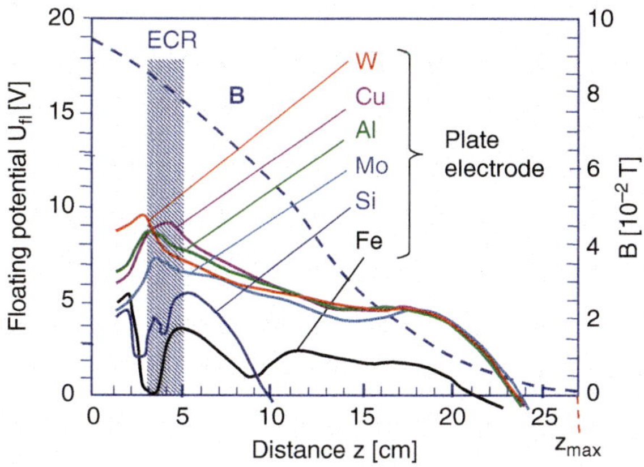


**Figure 4.24** Floating potential measured by a flat electrode plate with diameter of 40 mm made from different materials. The oxygen gas was admitted through the bottom of the discharge chamber and its pressure was $p_{O2}$ = 0.27 Pa.

values exceeding two, see Ref. [4.36]. On the contrary, the iron oxide ($Fe_3O_4$) has a lower maximum value $\gamma$ than pure iron, about 0.4–0.6, see Ref. [4.37]. The flat probe measured large radial areas, and the signals were affected by the peripheral parts of the plasma column. At lower plasma densities and larger distances (z), the measured signal reached also negative values and prevailing effect of electrons.

## 4.2 PE CVD of Silicon Nitride Films in the Far Afterglow

The processing of substrates in the decaying plasma, like the oxidation in an oxygen plasma introduced above, depends on the spatial distribution of the plasma parameters above the substrate. If the plasma is non-uniform with respect to the substrate surface, the resulting process (film deposition, oxidation or nitridation, etching, cleaning, plasma-ashing, etc.) will be non-uniform as well. Although the large-diameter plasmas would be naturally advantageous for uniform processes at smaller-diameters or small-diameter substrates, the basic arrangements for generation of the microwave plasmas (shown in Figure 2.11 in Chapter 2 and Figure 4.1 in this Chapter) can often suffer from the limited diameters of the discharge tubes. One example of the problem solution for the substrates with large surfaces is utilization of the space-resolved far afterglow (see Figure 3.5, Chapter 3). Such a method is based on the production and transport of the long-living metastable species generated by the plasma far from the processing area. These active particles can be distributed over large processing areas containing the samples and can react with suitable reactants to create homogeneous large-area processes. We will describe below an example of the high-quality silicon nitride films deposited by PE CVD process on the 20 cm diameter substrate table, see also Refs. [4.38–4.40].

The first tests were made in a small laboratory arrangement with the coaxial Surfatron (shown in Figure 2.11f, Chapter 2) for the microwave generation of nitrogen plasma at P ≤ 200 W in a quartz tube and the plasma-produced species divided into two inlets on sides of the glass reactor. An axial inlet to the reactor was terminated by a glass shower made from the fritted porous glass for diffuse delivery of the mixture 3% silane ($SiH_4$) diluted in Argon (Ar). The chamber pressure was between 0.5 and 1 Torr, controlled by the pumping speed. Figure 4.25 shows this arrangement.

The Surfatrons were designed in 1970 by Michel Moisan and coworkers, see Refs. [4.41–4.43]. The coaxial Surfatron is a tunable coaxial cavity terminated by an outlet slit launching surface waves propagating in the axis and losing their power along the discharge tube. Surfatrons can generate long plasma slabs in a broad interval of frequencies with low reflected powers (<5%). The plasma can be generated without or with an external magnetic field in a broad range of pressures from $10^{-3}$ Pa ($7.5 \times 10^{-5}$ Torr) up to the atmospheric pressure. These properties make the Surfatron systems flexible and versatile, e.g. in the plasma generation at moderately reduced pressures accessible with simple mechanical pumps.

The low-power arrangement showed in Figure 4.25 revealed that, after switching on the plasma, the area around substrate became luminescent and the sample surface was slowly coated by a silicon nitride film. However, due to low microwave power (limited by the coaxial power cable), the film growth was slow. Moreover, the reaction area was limited by the 8.5 cm diameter glass reactor, which limited the size of the samples. Therefore, the system was upgraded to a 30 cm diameter cylindrical steel reactor, and the Surfatron was arranged for the high microwave powers by a water-cooled coaxial coupled

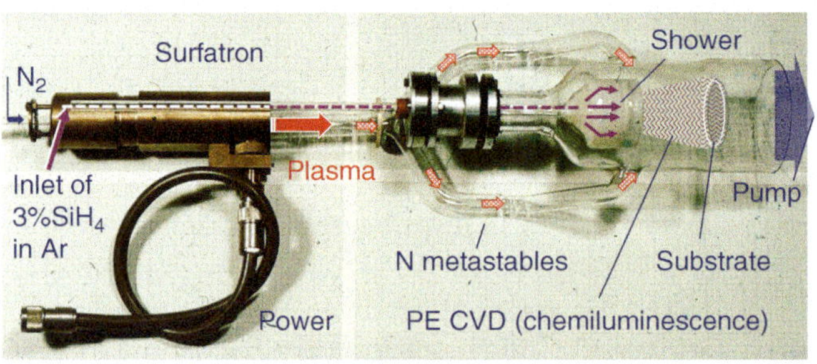


**Figure 4.25** Laboratory arrangement for PE CVD of $Si_3N_4$ films on substrates positioned outside the plasma region.

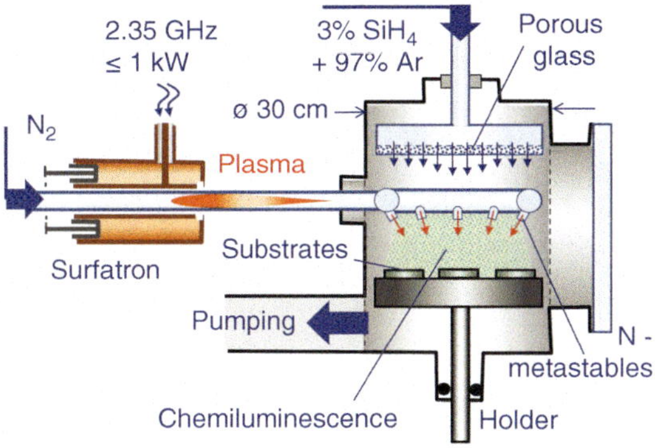


**Figure 4.26** Schematic of the experimental arrangement for PE CVD of silicon nitride films on large-area substrates.

with the rectangular waveguide. The schematic picture of this arrangement for deposition tests on substrates up to 20 cm in diameter is shown in Figure 4.26.

Figure 4.27 shows the whole arrangement and the view into the 30 cm diameter steel reactor. Inside is the circumferential glass shower delivering an activated nitrogen (N*) and a planar inlet of the diluted silane using a fritted glass in an Al frame. This reactor was tested for industry applications, but it was still not enough for large sizes of substrates and lower deposition rates as compared to those required in mass production.

The processes of PE CVD of the Si-N films in this far afterglow reactor were studied by an optical emission spectroscopy (OES) through the reactor window. A weak background emission from the remote nitrogen plasma generated by the Surfatron was first detected without inletting the silane mixture.

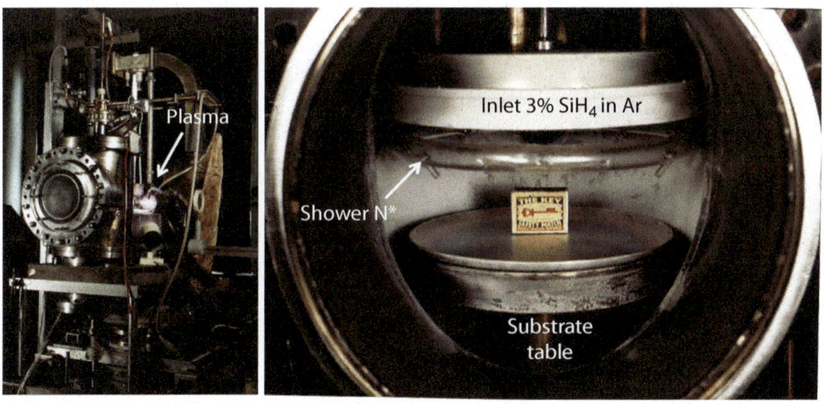


**Figure 4.27** The system (left) operating with the Surfatron-generated nitrogen plasma. A view inside the reactor with 20 cm diameter substrate table (right). The substrate table was equipped with an ohmic heating and a water cooling system for temperature control.

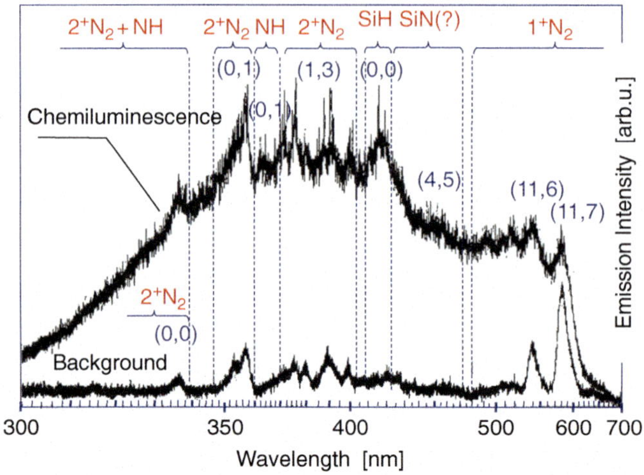


**Figure 4.28** Comparison of the optical emission intensities of the background from the remote nitrogen plasma and a chemiluminescence from the chemical reactions between the $SiH_4$ and active nitrogen particles from the remote plasma.

After the silane (3% in Ar) was inflowed, an intense emission of a yellow-green chemiluminescence was observed. The detected optical emission spectra before (background) and after silane inflow are shown in Figure 4.28.

The optical emissions (identified according to Ref. [4.44]) revealed an important role of the First Positive System of nitrogen molecules ($1^+N_2$) in forming of radicals of SiN, SiH, and NH. Further confirmation of the involvement of these radicals in the PE CVD process was found by measurements of the power and pressure functions of the relevant optical intensities of the heads of the most intense molecular bands in the emission spectra and their comparisons with the same functions of the growth rate of the Si-N film. These results are shown in Figures 4.29 and 4.30.

These comparisons indicated that the growth of the Si-N films was connected with the $1^+N_2$ molecules and radicals SiN and SiH. The additional studies confirmed the presence of significant density of the nitrogen atoms N ($^4S$) in the afterglow region, formed from $1^+N_2$, see Ref. [4.39]. Therefore, PE CVD process is based on the following reactions with the N($^4$S), SiH and SiN radicals, see Ref. [4.40]:

$$\begin{aligned} &6\text{SiH} + 8\ \text{N}(^4\text{S}) - \text{ surface} \rightarrow 2\ \text{Si}_3\text{N}_4 + 3\text{H}_2 \\ &3\text{SiN} + \ \text{N}(^4\text{S}) - \text{ surface} \rightarrow \ \text{Si}_3\text{N}_4 \\ &\text{SiN} + \ \text{H}_2 - \text{ volume} \rightarrow \ \text{SiH} + \ \text{NH} \end{aligned} \tag{4.3}$$

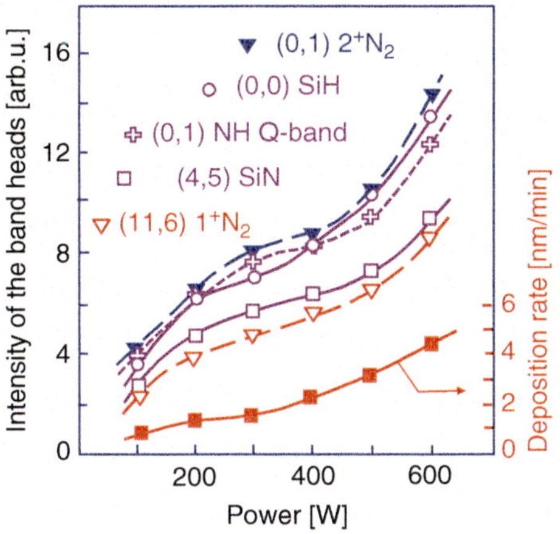


**Figure 4.29** Comparison of the power functions of the intensities of the most intense band heads and the growth rate of the silicon nitride. Total pressure $p_{TOT}$ = 80 Pa (0.6 Torr). Substrate temperature $T_{sub}$ = 300°C.

An additional support to the considered reactions was a visibly longer outreach of the $1^+N_2$ molecules compared to the $2^+N_2$ molecules, as can be seen in Figure 4.31. This figure shows a photograph of the nitrogen plasma column generated by the microwave Surfatron. The length of the column corresponds with the displayed length of the emission of $1^+N_2$ molecules.

Detailed tests of the properties of the afterglow grown Si-N films proved a high-quality stoichiometric $Si_3N_4$, see Ref. [4.45]. The films had the dielectric strength of up to $10^7$ V/cm, resistivity up to $10^{17}$ Ohm cm and a refractive index between 1.85 and 1.92. The values of these properties can be affected and controlled by the process parameters, e.g. by the microwave power, gas

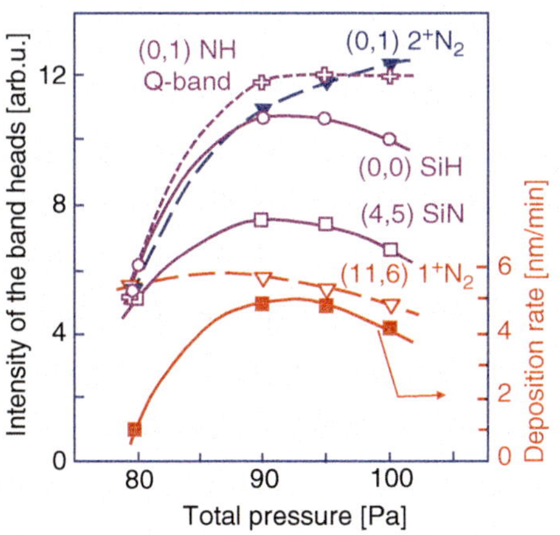


**Figure 4.30** Comparison of the total gas pressure functions of intensities of the most intense band heads and the growth rate of the silicon nitride. The microwave power P = 500 W. Substrate temperature $T_{sub}$ = 300°C.

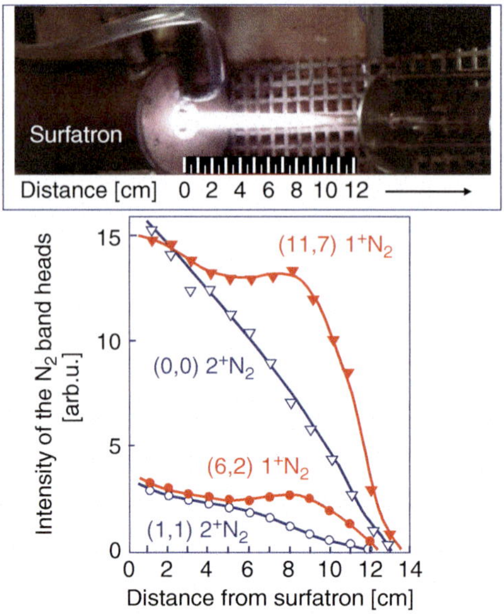


**Figure 4.31** The comparison of the optical emission intensities of the band heads of $1^{+}N_2$ and $2^{+}N_2$ molecules along the plasma column generated by the Surfatron.

flows, gas pressure, and sample temperature. Due to absence of the plasma in the PE CVD region and the low substrate temperatures, it was possible to deposit $Si_3N_4$ films directly onto the final semiconductor components without affecting their functions and properties. An example was a dielectric coating on power thyristors, see Figure 4.32. The far afterglow PE CVD systems confirmed possibility of quite large volumes and large-area depositions. To increase the film deposition rates further, such systems could utilize several generators of the microwave plasma delivering long-lived metastable particles, e.g. $N(^4S)$ atoms with uniform mixing of these particles with suitable gaseous precursors. Even though the growth rates of films in such systems might be still low (<100 nm/min), the process is well controllable and offers good film parameters. A similar process is not feasible for oxides because of short lifetimes of excited O* atoms.

Another interesting and important observation was made in the silicon nitride PE CVD. After the long operation of the deposition system, the quartz tube in the decaying part of the plasma and the whole inner part of the reactor were coated by an amorphous whitish powder, see Refs. [4.31, 4.46]. A consequence of this was that the growth rate of films was substantially reduced, see Figure 4.33 (for sccm units, see the Appendix). An infrared analysis showed this white powder contained nitrogen, hydrogen, and silicon bonds, with a prevailing density of N-H bonds. It indicated the transport tube walls were coated by PE CVD products containing a certain density of N-H bonds and

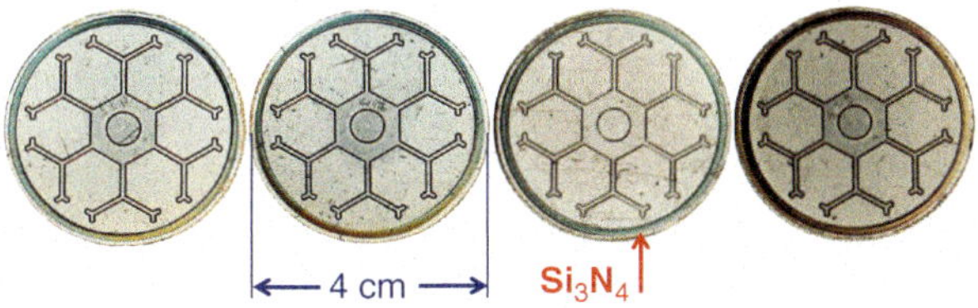


**Figure 4.32** A set of the power thyristors coated by $Si_3N_4$ films in the far afterglow PE CVD.

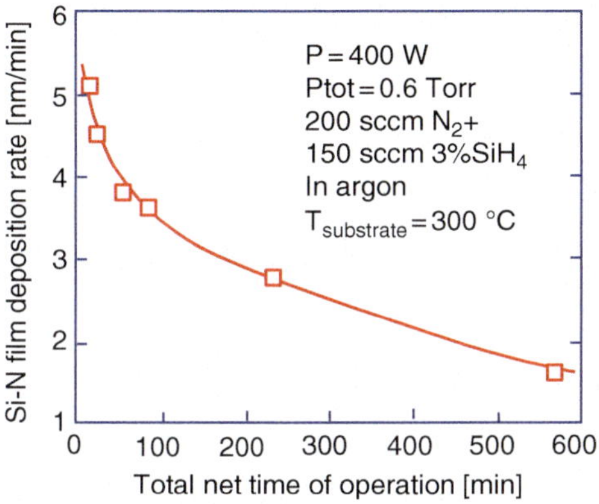


**Figure 4.33** The time-dependency of the far afterglow PE CVD rate of Si-N films.

these could effectively absorb the active nitrogen atoms necessary for the nitride film growth. The successive decrease of the film growth rate was observed also by other authors in the PE CVD of the Si-N films, see Ref. [4.47], and the Si-O films, see Ref. [4.48]. This effect evidently relates with the successive deposition of coatings absorbing the active species. All plasma reactors exhibit some kind of a memory based on a fouling of the walls, which affects the rates of the processes either after changing the process components or after long operation times. A periodic cleaning of the reactor is general requirement in all plasma processing practices. The cleaning (e.g. by a weak HF acid) was successful also in this case.

## 4.3 Microwave Plasma Jets for PE CVD of Films

The plasma processing is often needed on small areas or inside narrow profiles and tubes. The conventional microwave plasma systems are not useful, mainly because of the relatively high power (order of 0.1 to 1 kW) needed for generation and sustaining the plasma. Moreover, transport of the microwave power into a reactor through dielectric windows is often accompanied by undesirable coatings on the window or even its damage. For example, hydrogen plasma on quartz can cause a reduction of the oxygen from silicon dioxide ($SiO_2$) and result in an Si coating. As the silicon (Si) coating absorbs the microwave power, it increases the wall temperature, thereby enhancing the process and the local melting of the tube with the consequent vacuum leak. This can easily happen at higher microwave powers. Therefore, all windows almost always represent a weak part in the conventional microwave plasma systems.

To avoid problems with the windows a novel reactor was designed using a plasma slab generated by waveguide Surfatron as an antenna "electrode", or by installing an auxiliary metal antenna electrode in the axis of the Surfatron, see Ref. [4.49]. Such system required low microwave power (order of only 10 W in most cases) for generation of a plasma slab or a plasma ball close to the substrate. For testing of a broad range of plasma parameters, the gas tube with the plasma antenna or the metal antenna had adjustable lengths. This enabled one to optimize the antenna length for the resonance regimes or to work in a downstream regime, where the antenna acted as a microwave plasma jet for local film depositions.

The movable plasma antenna, i.e. the plasma slab can be narrow and can be applied inside narrow holes for processing purposes. Figure 4.34 shows about 2 mm diameter plasma slab immersed in 6 mm diameter hole with 20 mm depth. In PE CVD applications, a slow shifting of the antenna in the hole can provide satisfactory uniform thickness of the film deposited inside. The described arrangement can be used to emphasize the bulk plasma processes by tuning the resonance in the reactor or to emphasize a narrow plasma slab in the plasma antenna for processing inside the hollow parts of the substrate. In this way, it was possible to plasma-treat complicated surfaces in one processing cycle.

Figure 4.35 shows the PE CVD test system with an adjustable length of the plasma antenna generated by the waveguide Surfatron (called the Surfaguide)

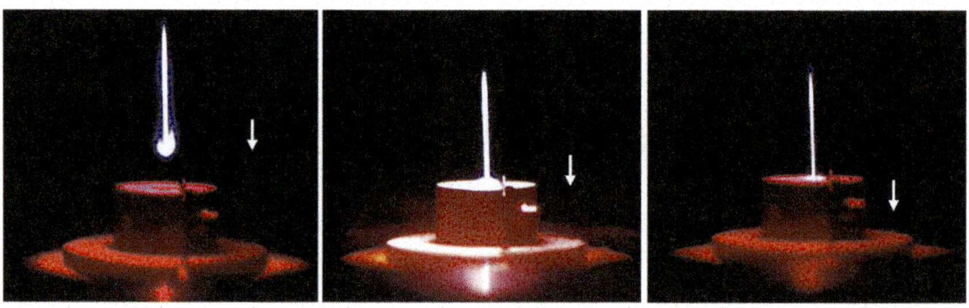

**Figure 4.34** The 2 mm diameter argon plasma slab generated by 10 W microwave power can be immersed into 6 mm diameter 20 mm deep hole.

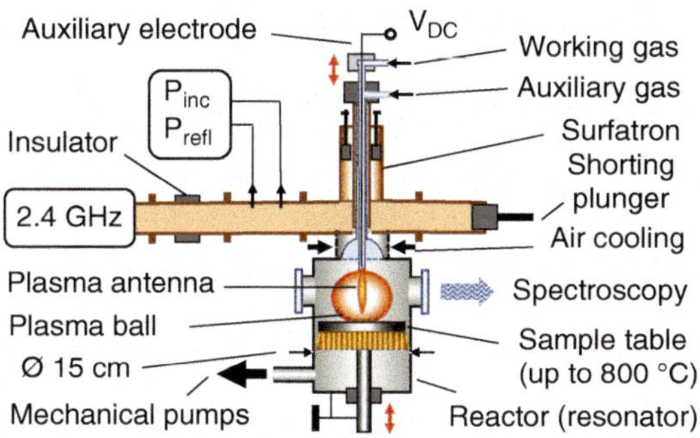


**Figure 4.35** The PE CVD system with the plasma antenna generated by the waveguide Surfatron. The working gas is fed inside or around the antenna and an auxiliary gas around the central tube. The movable sample table tuned the resonance and has controllable temperature.

in a tunable reactor (resonator) called the Surfajet or electrode microwave discharge (EMD) or microwave antenna discharge (MWAD). The used steel reactor (15 cm in inner diameter) served as a microwave resonator. It was equipped with water-cooled walls and a movable short for adjusting the resonance length and holding the substrates during processing. This movable short had controlled temperature by means of water cooling or by ohmic heating up to 800°C. The photograph of this shorting plunger is shown in Figure 4.36.

The resonance in the reactor required the proper length (l) of the antenna and the length (L) of the reactor vessel adjusted by the movable short, see Figure 4.37. The resonance position of the movable shorting plunger in the reactor was identified by the steep decrease of the reflected microwave power.

The Surfajet system was tested first for the PE CVD of diamond films, see Refs. [4.50, 4.51] and then for the $CN_x$ films, see Refs. [4.52, 4.53, 4.54]. For diamond

**Figure 4.36** Detail of the movable short for the reactor with the central holder of samples. The inner diameter of the reactor was 15 cm and its side flanges were shielded by metal nets.

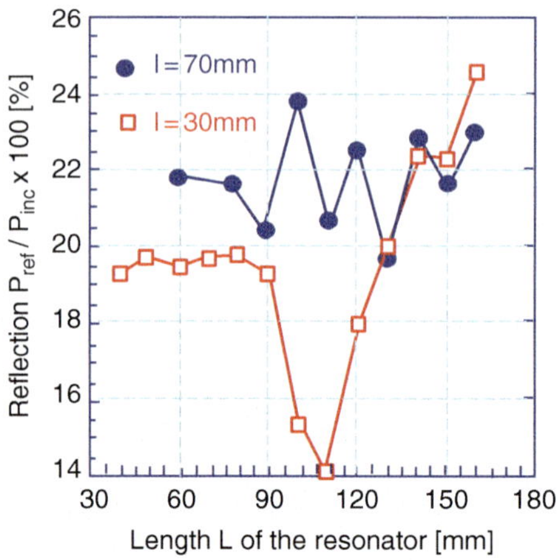


**Figure 4.37** Adjusting the reactor length L to resonance at two different lengths l of the antenna. Antenna gas was 300 sccm $H_2$ + 3sccm $CH_4$. Auxiliary gas was 100 sccm Ar + 1sccm $O_2$. The incident power was $P_{inc}$ = 400 W, the gas pressure in the reactor was $p_{TOT}$ = 5 Torr.

depositions, the gas mixtures composed of $H_2 + CH_4 + O_2 + Ar$ were examined. The films were deposited on Si and Mo substrates. Typical experimental parameters for the diamond deposition were: durations of 60–900 min, the incident microwave power $P_{inc} = 300–700$ W; the total reflection $P_{ref}/P_{inc} \leq 10\%$, table temperature $\leq 800°C$; and total gas flow 300–600 sccm. The tested gas compositions related to 100% $H_2$ were 0.5% – 2% $CH_4$, (< 0.5% $O_2$) and ≤ 50% Ar. Argon was used as a shielding gas. The total gas pressure measured close to the substrate was 2–10 Torr ($2.7 \times 10^2 – 1.3 \times 10^3$ Pa). The resulting diamond films are shown in Figure 4.38. All films exhibited sharp Raman peak at 1335 $cm^{-1}$.

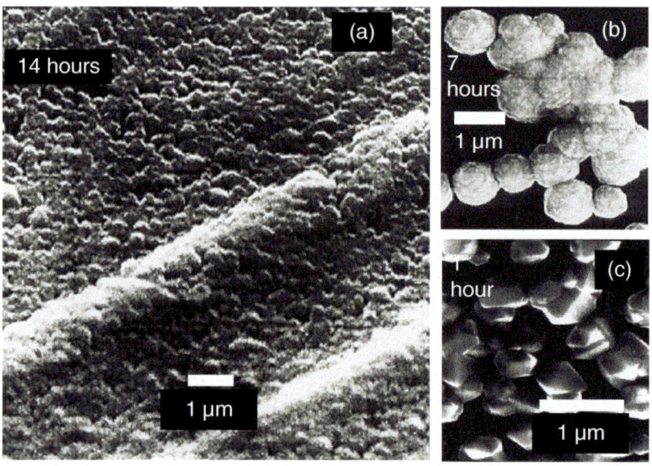


**Figure 4.38** (a) Diamond film deposited on Mo substrate in the resonant regime with plasma antenna. (b) Diamond crystals on Si substrate (400 W, 750°C, 5 Torr), Ref. [4.49]. (c) Diamond crystals created on Si substrate by PE CVD with metallic antenna (200 W, 750°C, 15 Torr), see also Ref. [4.51].

### 4.3.1 Deposition of Carbon Nitride Films

At the microwave power, as low as 10 W, the growth rate of the $CN_x$ films from nitrogen and acetylene, in the Surfajet reactor with the metallic antenna in the axis of the Surfaguide cavity (shown above in Figure 4.35), reached the order of 1 μm/min. This power was lower than the powers required in the PE CVD of diamond films. Because of the low power and the growth rate of C-N films was almost ten times faster than in most conventional methods (laser-assisted, ECR, or hot filament activation), such non-conventional microwave antenna system can be suitable for different small-scale industrial applications, e.g. in the microelectronic chip operations.

The axially installed microwave antenna was made from the stainless steel rod or tube, 6 mm in outer diameter and 4 mm inside. It served as an inlet of the $N_2 + C_2H_2$ gas mixture. However, for better film uniformity, the gas was often admitted around the antenna and not through it. The $N_2$ flow was 500 sccm or 250 sccm, mixed with different flows of $C_2H_2$ selected in the interval between 1 and 5 sccm. Figure 4.39 shows the photograph of the Surfajet system. The microwave tuning elements used in this system were the following: the movable coaxial plunger-tuner inside the Surfaguide, the waveguide short at the end of the power line, and the movable sample holder-plunger short in the reactor chamber (detailed view is in Figure 4.36 above).

The $CN_x$ films were deposited by the PE CVD for 10 minutes at the microwave power selected between 10 and 50 W. The microwave plasma generated in the mixture of nitrogen with acetylene had a typical spherical shape with a

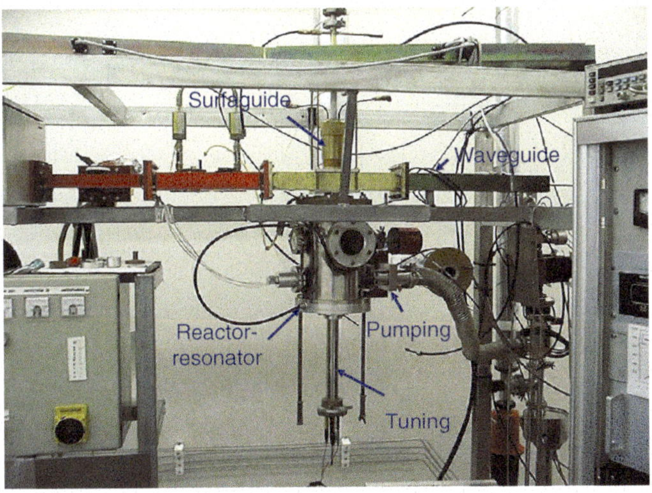


**Figure 4.39** The Surfajet system with 15 cm inner diameter reactor-resonator. The mechanical pumping system consisted of Roots blower and two-stage rotary pump.

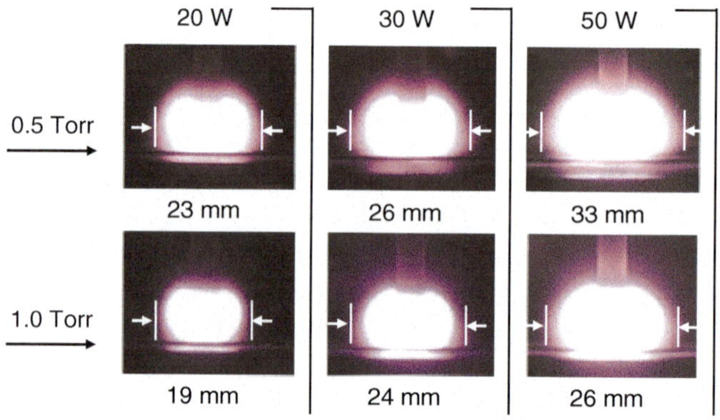


**Figure 4.40** Photographs of the ball plasma generated in $N_2 + C_2H_2$ gas at different microwave powers for two reactor pressures. Diameter of the plasma ball depends on both parameters.

diameter dependent on the dissipated power in the reactor and on the reactor pressure, see Figure 4.40. The plasma ball diameter decreased with the increasing gas pressure and increased with the increased microwave power. Diameters of films deposited on the Si samples, alternatively on paper or transparent plastic foils, were limited by an auxiliary steel mask to 20 or 30 mm

Detailed observations revealed that, at higher pressures, the plasma ball acquired more intense central part, which led to a thickness peak in the film center. Above a certain pressure, the maximum growth rate decreased. However, this depended on the applied microwave power, see Figure 4.41.

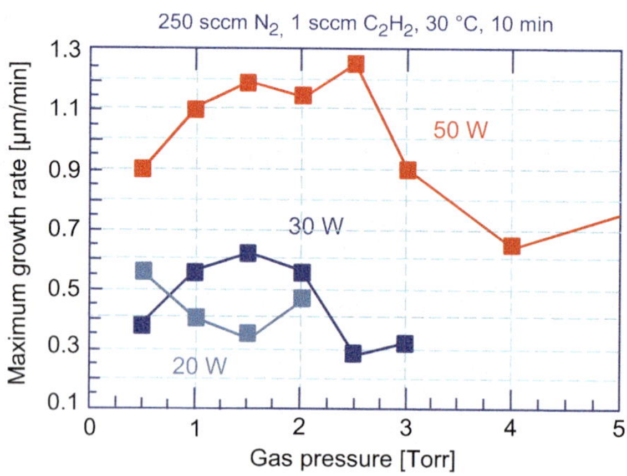


**Figure 4.41** Pressure dependence of the maximum growth rate of $CN_x$ films on Si substrates.

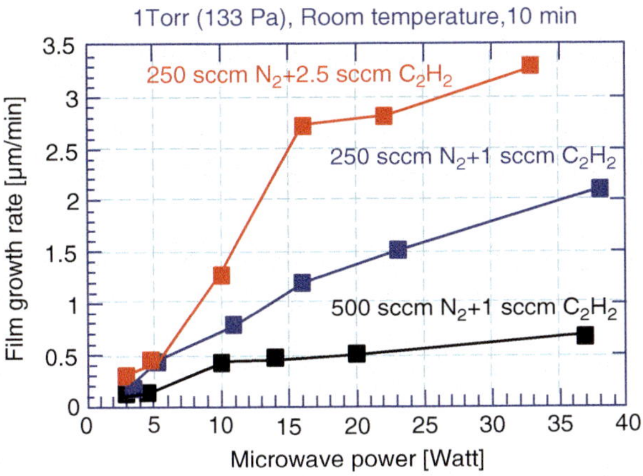


**Figure 4.42** Power dependence of the maximum growth rate of $CN_x$ films on Si substrates.

The maximum thickness in the center of the sample was observed at short distances of the sample from the antenna electrode or at gas pressures less than 1 Torr. The film growth rate always increased with the increased microwave power, see Figure 4.42.

Typical examples of the $CN_x$ films grown for 10 minutes on Si substrates at different temperatures are shown in Figure 4.43. The film grown at room temperature had poor adhesion.

The $CN_x$ films grown on silicon in the antenna-arranged Surfajet plasma in the $N_2 + C_2H_2$ mixtures had mostly fibrous structures with the nitrogen content of $x \approx 10\%$. The diameter of the fibers decreased with the deposition temperature, see Figure 4.44.

The structure of the $CN_x$ films on Si substrates differed from films on the Al foil substrates, see Figure 4.45.

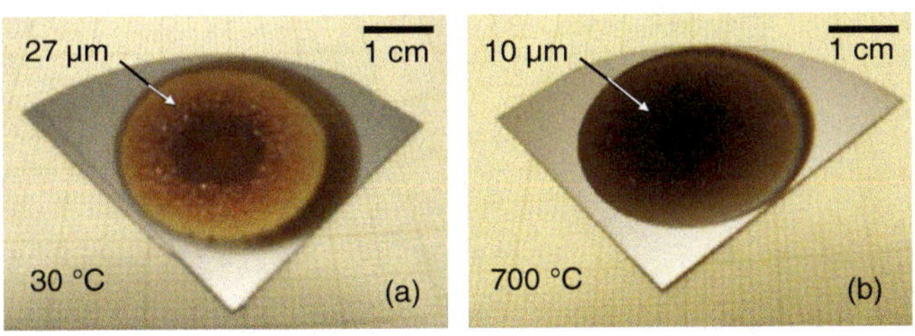


**Figure 4.43** An example of thick $CN_x$ films grown at 10 W microwave power for 10 minutes on Si substrate unheated (a) and heated to 700°C (b). Gas pressure was 1 Torr. Thick film grown at room temperature had high compressive stress and poor adhesion.

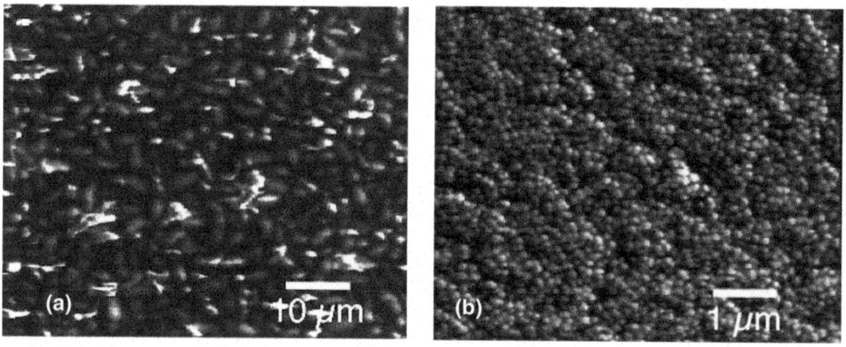


**Figure 4.44** The SEM images of fibrous structures of the $CN_x$ (x ≈ 10 at%) films grown by the microwave antenna PE CVD at 1 Torr (30 W, 2.4 GHz, 250 sccm $N_2$ + 2.5 sccm $C_2H_2$). Fiber diameter decreases from about 1-3 μm at 300°C (a) to about 80 nm at 550°C (b). Ref. [4.52].

The fibers were grown to diameters of almost 10 μm. Such thick columns had good adherence to the substrate, and they followed scratches on its surface. The low temperature of substrates during PE CVD allowed also coating of plastics and paper (see Section 4.3.2).

The measurements of the vibrational temperature ($T_v$ or $T_{vib}$) of $N_2$ molecules revealed that the chemical activity of the nitrogen-based reactive microwave

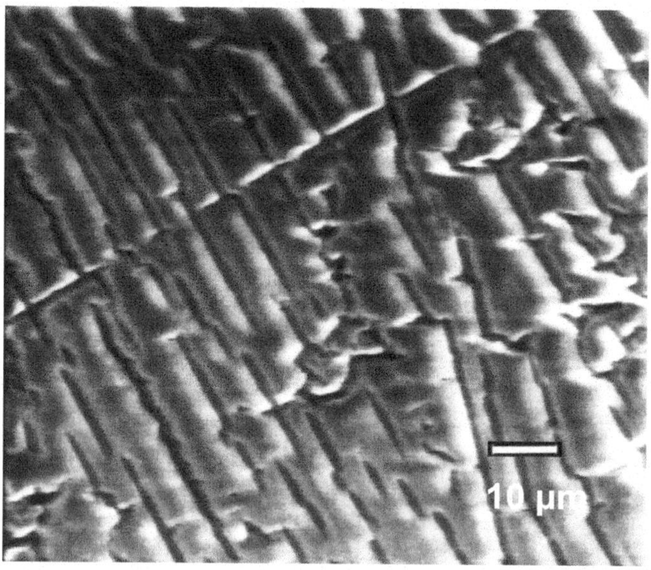


**Figure 4.45** An aluminum foil substrate coated 10 minute by the $CN_x$ film in the Surfajet reactor.

plasma depended on the vibrational temperature, which affected the rate of plasma chemical processes. For vibrational temperature see Chapter 2, Section 2.5.2. The $T_{vib}$ was estimated from the $N_2$ ($C^3\Pi_u$) band intensities, see Ref. [4.55]. The experimental error of the temperature values didn't exceed ± 200 K. Because the $C_2H_2$ content in nitrogen was low (≤1%), the radial profiles of $T_{vib}$ were calculated from the emission spectra of the $N_2 + C_2H_2$ plasma for two $N_2$ flows. The distribution of the vibrational temperature $T_{vib}$ across the plasma ball is shown in Figure 4.46, together with the radial profiles of the film thickness.

At lower nitrogen flowrates, a slight increase appeared in the vibrational temperature around the central part of the plasma. Even this tiny feature was followed by a visible increase of the film growth rate in the central part of the substrate.

### 4.3.2 Surfajet Plasma Parameters and an Arrangement for Expanding the Plasma Diameter

The Surfajet plasma generated with the metallic antenna in the axis of the Surfaguide was analyzed by the Langmuir probes (for probes, see Appendix). Some of the results were published, see Ref. [4.56]. The used double Langmuir

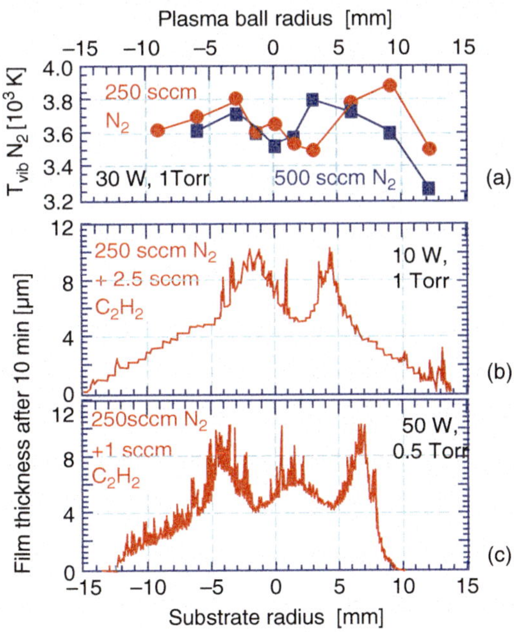


**Figure 4.46** Vibrational temperature of the $N_2$ ($C^3\Pi_u$) emission system measured across the plasma ball (a). Radial profile of the $CN_x$ film thickness after 10 minute PE CVD at 10 W and 1 Torr, (b) and at 50 W and 0.5 Torr, (c).

probe consisted of two parallel tungsten wires 100 μm in diameter with 1 mm long uninsulated parts, at 3 mm distance from each other. The glass insulation was used on the rest of the probes, and a conductive tube holding the system served as an electrical shielding. The probes were used for measurements of the saturated ion current (at −50 V DC) and an electron temperature estimated from the slope of the probe characteristics. The plasma was generated in hydrogen or in $H_2$ + $CH_4$, similar as with the deposition of diamond films. Because the probe can be affected by the microwave field, the measurements of the electron temperature were performed in the region outside the central plasma column. The measured values of the electron temperature were about 1–3 eV. For estimates of the plasma density using the expression in Eq. (4.2) in Section 4.1.2, identifying the contributing ions was necessry. In the hydrogen plasmas with the degree of dissociation below 50%, the most abundant ions are $H_3^+$ produced by the ion conversions. Considering the surface of the probes for the collecting of ions in the saturated ion current $I_s$ the hydrogen plasma density was estimated to be about 2 $10^{11}$ $cm^{-3}$ in the center of the system, with an exponential drop in the radial direction, see Ref. [4.56].

Further studies of the Surfajet system confirmed that expanding the diameter of the plasma ball by conical termination of the antenna is possible. Figure 4.47 shows the schematic illustration of the Surfajet arrangements with the cylindrical antenna and the conical antenna. The figure shows position of the movable Langmuir probe about 4 mm above the surface of the substrate. It is also shows how the individual gases were inflowed into the reactor. The arrangement illustrated in Figure 4.47b had a 20 mm diameter bottom of the conical antenna. It was tested for the diamond films at high substrate temperatures, higher powers and for the low-power PE CVD depositions of the $CN_x$ films.

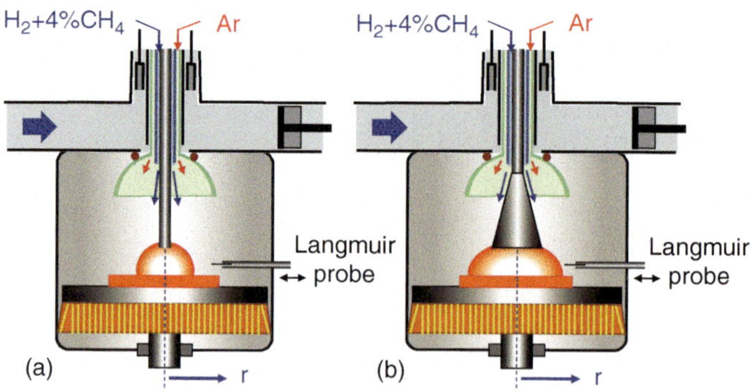


**Figure 4.47** Comparison of the cylindrical antenna (a) and a conical antenna (b) in the Surfajet reactor. The reactor was arranged with the Langmuir probe for the parameter measurements.

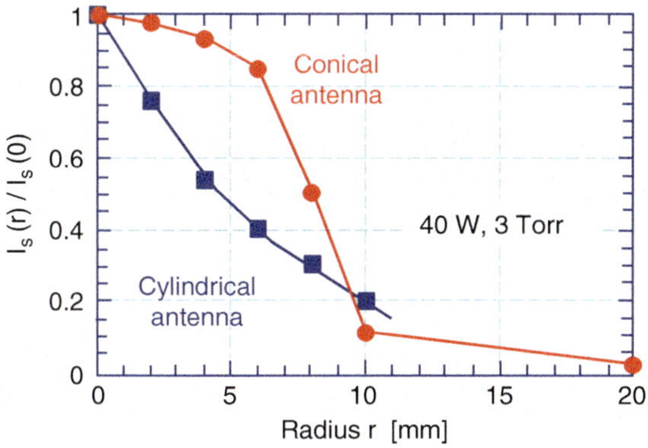


**Figure 4.48** Comparison of the normalized saturated ion currents in the plasma of $H_2$ + $CH_4$ for the cylindrical and conical microwave antenna.

The measurements of the saturated ion current in the gas mixture of $H_2$ + $CH_4$ showed that, at an identical power, the conical antenna was capable to expand the plasma density to larger radius, which enabled depositions on larger substrate diameters. Figure 4.48 shows the comparison of the saturated ion current distributions for cylindrical and conical antennas.

Figure 4.49 shows the comparison of the real discharges in the mixture of $N_2$ + $C_2H_2$. The expanding of the plasma diameter at identical microwave power is visible.

As an example, the filter paper substrate coated by the $CN_x$ film at an area of 30 mm in diameter in the Surfajet device with the conical antenna of 20 mm

**Figure 4.49** Surfajet discharges in $N_2$ + $C_2H_2$ at microwave power of 50 W and 1 Torr pressure.

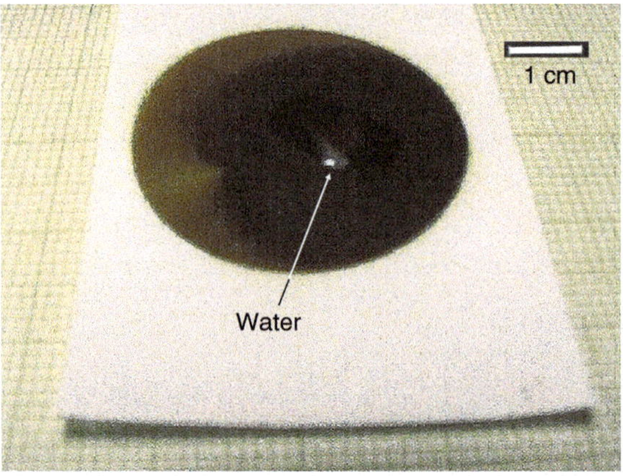


**Figure 4.50** A filter-paper substrate coated 10 min by the $CN_x$ film on a 30 mm diameter area was capable of keeping the water droplet.

in diameter was capable to hold the water droplet, see the photograph in Figure 4.50.

It is possible to conclude that the described non-conventional arrangements used in the Surfajet reactor could be useful for different PE CVD applications on small components.

## 4.4 Hybrid Microwave Plasma System with Magnetized Hollow Cathode

An original hybrid plasma source (called HYP) combined the microwave ECR plasma with the magnetized hollow cathode was designed as a system sharing the common magnetic field. The microwave ECR systems generate high-density plasmas, but due to the high frequency, the main energy receivers are electrons. On the contrary, the RF or DC generated hollow cathodes generate the plasma with high energy ions. Therefore, a synergetic combination of both sources enabled a controllable process in the plasma density and the ion energy for the plasma processing. A schematic illustration of the hybrid source (HYP) designed, developed and tested by the authors is in Figure 4.51, see Ref. [4.57].

The HYP source combined an ECR microwave plasma generated in a horn waveguide antenna by the magnetic field with the ECR induction level positioned at the outlet slit of an RF-powered parallel-plate hollow cathode installed inside the antenna horn. The hollow cathodes of several forms,

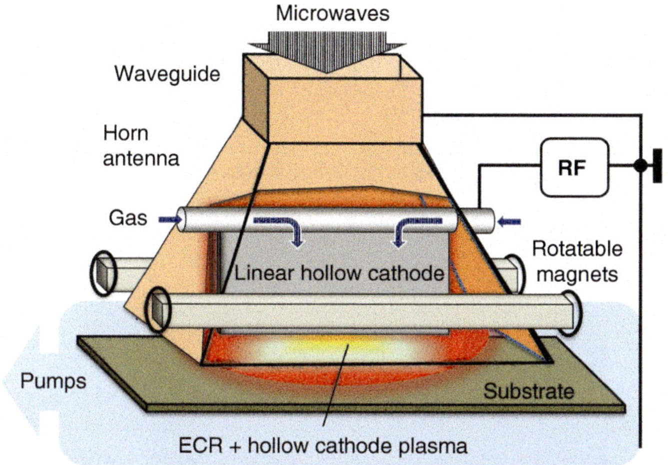


**Figure 4.51** Schematic illustration of the hybrid plasma source (HYP) combining the ECR microwave plasma with the magnetized M-M hollow cathode source with a pair of the cathode plates powered by the radio frequency generator.

particularly those generated by the radio frequency (RF) or a pulsed DC power, have been described in many applications, see, e.g. Refs. [4.58–4.61]. The linear parallel-plate hollow cathode operated in the magnets-in-motion (M-M) arrangement in the HYP source used rotatable magnets with periodic changes of the magnetic polarization across the outlet slit of this linear cathode. The photograph in Figure 4.52 shows the front view of the HYP source with the waveguide horn antenna coupled to 2.4 GHz power (up to 2 kW) and the coaxial cable connecting the RF generator 13.56 MHz (up to 3 kW) to one side of the M-M hollow cathode with 16 cm long slit between cathode plates installed in the horn antenna.

The ECR magnetic field (≈ 875 G) was generated by two opposite rotatable strong permanent magnets (NdFeB) positioned at the level of the outlet slit of the M-M hollow cathode. The microwave power was applied by the rectangular waveguide to the horn antenna through the special sealed ceramic window. Such a window was described in Figure 1.18 (Section 1.4.5, Chapter 1). The whole assembly with the antenna and the linear M-M hollow cathode was pumped by simple vacuum system composed of the diffusion pump 700 l/s (Diffstak 160/700 P) backed by a Roots blower coupled with the 80 $m^3$/h rotary pump. The working gas (Ar, $N_2$ or their mixture) was introduced between the parallel plates of the hollow cathode. The hollow cathode was powered from the RF generator through an automatic matching unit. Figure 4.53 shows two modes of the plasma in the HYP reactor: the ECR microwave plasma alone and

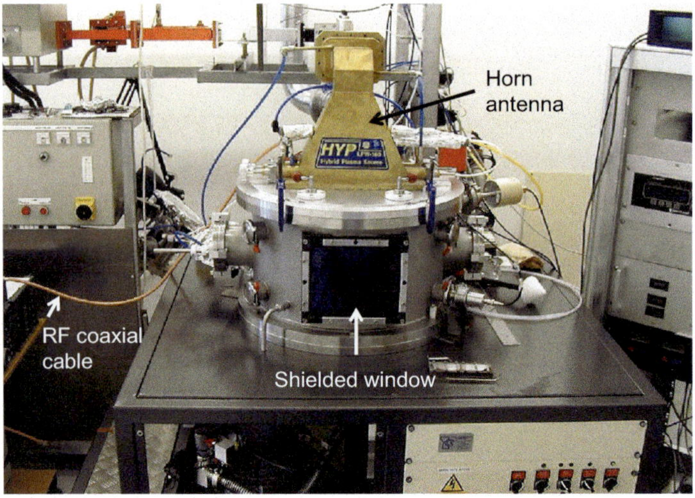


**Figure 4.52** The front view of the HYP source for the 2.4 GHz microwave power and 13.56 MHz RF power. The vacuum chamber has 450 mm of the inside diameter and the water-cooled walls.

the hybrid plasma consisted of the microwave ECR plasma with the RF generated hollow cathode plasma. To protect the microwave radiation from the reactor, the window was shielded by a metallic net.

The plasma parameters in the HYP reactor were measured by a movable double Langmuir probe installed 15 cm below the outlet of the M-M cathode to be far from the strong magnets. Part of the results was published, see Ref.

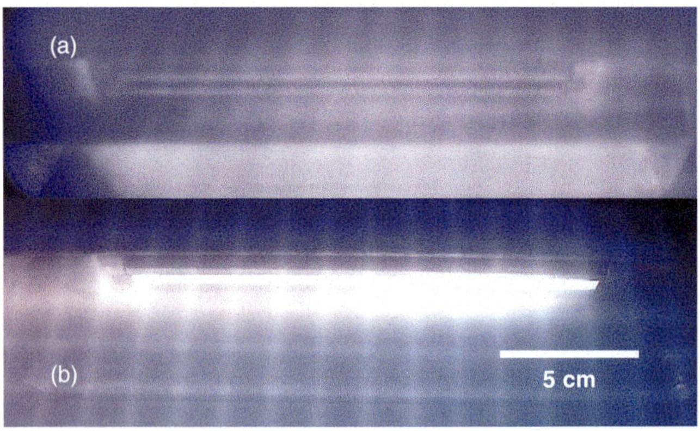


**Figure 4.53** View on the argon plasma at 6 mTorr (0.8 Pa) pressure in the HYP source through the window shielded by a metallic shielding net. (a) The ECR plasma at 1 kW. (b) The ECR plasma with the RF hollow cathode plasma generated by the M-M source at 1 kW power.

[4.62]. Figure 4.54 shows the results of the Langmuir probe measurements of the plasma density (ion density = electron density) and the electron temperature generated by individual plasma sources in the central part of the HYP reactor 15 cm from the hollow cathode outlet, in different mixing of $N_2$ with Ar. The hollow cathode can generate higher plasma density because it produces focused stream of the plasma while the ECR plasma is being distributed in the volume. The plasma density, generated simultaneously by both sources, could reach over $6 \times 10^{11}$ $cm^{-3}$ at 1 kW of the incident powers delivered from the RF and the microwave generator, in pure argon at chamber pressure of 6 mTorr (0.8 Pa). Accounting for the reactor volume of about 60 $dm^3$ (60 liters), this density classifies the HYP source as the high-density plasma source.

The HYP source was found to be efficient in the depositions of films. The reactive PVD was tested for TiN and CrN films, and the PE CVD regime was tested for the carbon-based films. The presence of the microwave power is beneficial for enhancement of the ionization, but it can affect conditions of the substrate itself during the deposition. As explained in Chapter 1, Section 1.4.7, Table 1, the microwave power can be absorbed efficiently in the silicon substrates and can slightly increase the surface temperature. This improves the coated TiN film adhesion and a tightness of the interface between the Si surface and the film.

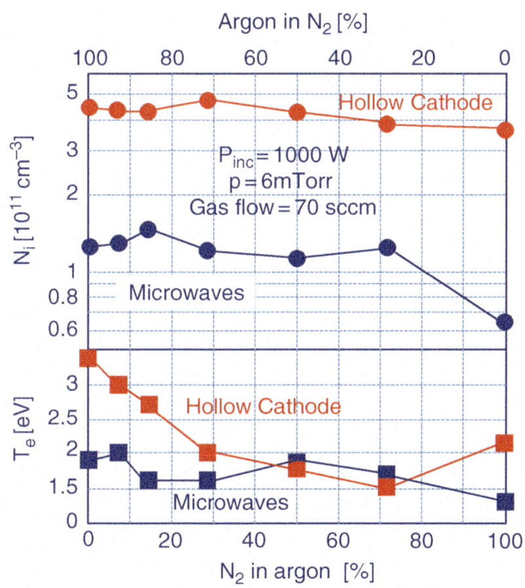


**Figure 4.54** The ion density and electron temperature measured by double Langmuir probe 15 cm below the hollow cathode outlet for different mixtures of argon with nitrogen.

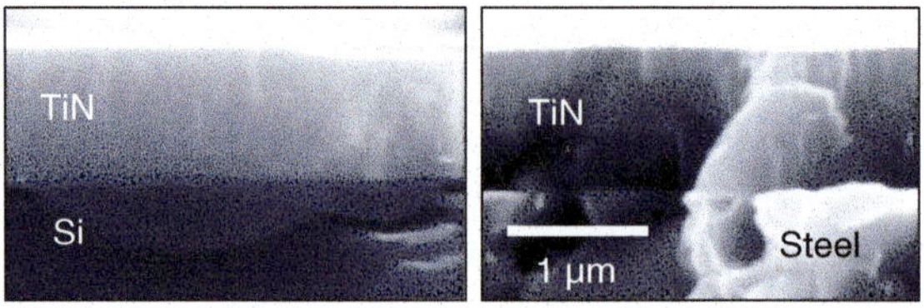


**Figure 4.55** The scanning electron microscope (SEM) image of the fracture cross-sections of the 1 μm thick TiN films on silicon and on stainless steel substrate.

Examples of the 1 μm thick and highly (111) textured TiN films deposited on silicon and stainless steel substrates deposited in the HYP source are shown in Figure 4.55. Typical texture of the films was almost always in the "zone T" of the Thorton's texture model for the sputtered PVD films, see Ref. [4.63].

Figure 4.56 shows the depth profiles of the elemental atomic contents measured in the deposited films. The profiles clearly confirmed the stoichiometric TiN film, i.e. 50% Ti and 50% N. Typical growth rates of these films in the HYP source were between 0.1 and 0.3 μm/min, which is about 1.4 times higher than in the magnetron sputtering systems under comparable conditions. Moreover, in the reactive PVD of the CrN films, the obtained growth rates were 4 μm/min, which are inaccessible in the magnetron sputtering devices.

Deposition processes in the low-pressure microwave systems, like the HYP source, can coat the microwave power inlet window and cause lowering of the delivered power or crash of the window. Therefore, preventing or slowing the coating on the window was necessry. A simple solution was an "S" profile of the waveguide connected to the horn antenna chamber. Figure 4.57 shows this solution.

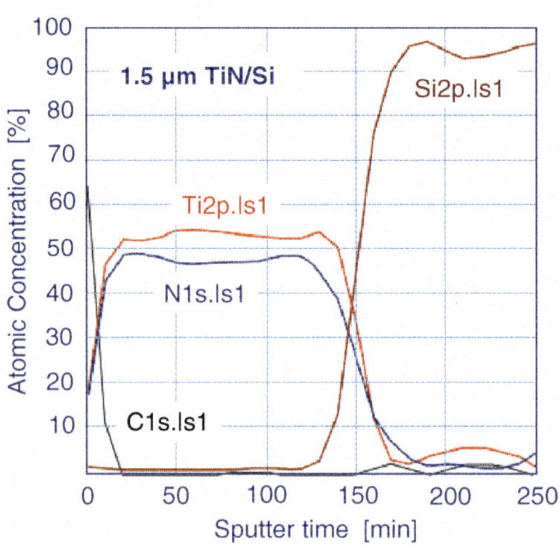


**Figure 4.56** The ESCA depth profile in the 1.5 μm thick TiN film deposited 10 min by the reactive PVD on Si substrate in the HYP source. The process parameters: 6 mTorr, 70 sccm (Ar + 10% $N_2$), RF power 3 KW, μW power 1.2 kW.

**Figure 4.57** Vacuum chamber (horn antenna) of the HYP source with the "S" profile on the waveguide inlet. The bottom flange of the chamber is arranged for the rotatable magnets.

Another important observation was made during the HYP source coating on the ferromagnetic steel substrates. When the HYP source was used for the PVD of TiN films on ordinary steel substrates, the directions of ion fluxes in strong magnetic fields were affected by changes of the field geometry and attracted to the substrates. The ferromagnetic material of the substrate attracted the magnetic power lines to its surface and enhanced the ion bombardment. This was not observed in case of the diamagnetic samples, including austenitic steels. The effect was confirmed on the steel components made as a combination of the austenite (diamagnetic) steel and the martensite (ferromagnetic) steel. Deposition of such a component revealed different film quality (adhesion, roughness), better on the austenite surface than on the martensite surface. Such effect is of a more general validity. It was tested and verified in the conventional magnetron sputtering, see Ref. [4.64].

The HYP source deposition experiments on the silicon and steel samples revealed differences in the sample temperature measured during PVD. The higher temperature of the silicon samples can be caused by the absorption of some part of the microwave power not consumed for the generation of the plasma. In addition, it was confirmed that the PVD process can be affected by the substrate bias, but due to different electric resistances of the silicon and the steel, the resulting ion bombardment was different, see Ref. [4.65]. Therefore, the optimal PVD conditions must always be tuned to the individual substrates to be coated.

In summary, the advantages of the HYP sources for coating processes were the following:

- High or very high maximum film deposition rates.
- Ion-assisted film growth due to high-density plasma, which enables excellent films.
- Microwave radiation-assisted growth affecting several growth mechanisms, e.g. the process chemistry, surface temperature.
- A possibility of fast and efficient PE CVD regimes.
- Unique possibility to combine the PVD and PE CVD regimes.

Some drawbacks found in the tested arrangement of the HYP source were the following:

- A complex system, which required simultaneous operation of two power sources.
- A small target area producing metal particles in the PVD regimes (a narrow slit-type target, the cathode length is limited by the size of the microwave horn antenna).
- High cathode temperature necessary for the fastest arc-assisted PVD regimes.

Simple estimates accounting the operation parameters used in the HYP source confirm possibility of upgrading the system for industrial scale coating of multiple components, like engine valves, drills, or small tools. A simple design of such a device with two HYP sources, in Figure 4.58 confirms feasibility of an industrial usage of the HYP sources in mass production.

Based on numerous experiments with the HYP source, evaluating the advantages of the industrial coating device is possible, for example on principal design of an industrial device shown in Figure 4.58. The length of the hollow cathodes in the device can be about 16 cm and the deposition area will be about 3 cm wide. The typical PVD rate for hard TiN films is 0.2 μm/min; for hard CrN films, it is over 1 μm/min. For the PE CVD of hard carbon films (DLC = diamond like carbon), the rate can be about 5 μm/min. Reliable estimates of the production time, including pumping time for the flat components with about 3 cm in diameter, give about 200 pieces coated per hour at the estimated cost per piece of less than 0.025 EUR. Because of good properties confirmed in the coatings obtained by the HYP devices, a high probability exists for successful utilization of such non-conventional devices in the industry with a short time of the investments return for the building and installation. An additional option is for example application of 24 cm microwave systems (1.25 GHz), with rectangular waveguide (WR770) having inner dimensions roughly a ≈ 196 mm and b ≈ 98 mm, at about two times higher microwave power and with double

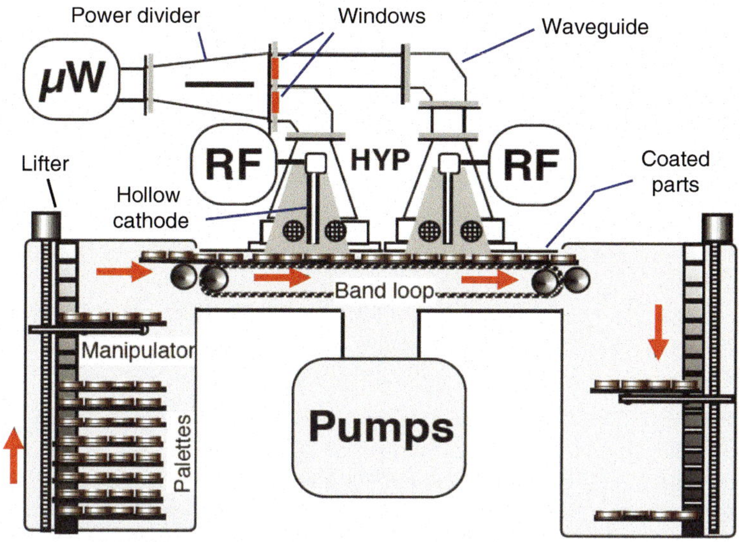


**Figure 4.58** A principal design of an automatic industrial device for coating of planar components with two HYP sources. The microwave power should be ⩾ 2 kW and each RF power should be ≈ 3 kW.

lengths of the linear hollow cathodes inside. However, these ideas remain theoretical.

## References

[4.1] J.R. Ligenza: "Silicon oxidation in an oxygen plasma excited by microwaves"; J. Appl. Phys. **36** (1965) 2703–2710.

[4.2] O.A. Weinreich: "Oxide films grown on GaAs in an oxygen plasma"; J. Appl. Phys. **37** (1966) 2924.

[4.3] J. Kraitchman: "Silicon oxide films grown in a microwave discharge"; J. Appl. Phys. **38** (11) (1967) 4323–4330.

[4.4] F.C. Fehsenfeld, K.M. Evenson, and H.P. Broida: "Microwave discharge cavities operating at 2450 MHz"; Rev. Sci. Instrum. **36** (3) (1965) 294–298.

[4.5] G. Lisitano: Proceedings 7th International Conference Ionization Phenomena in Gases, Gradevinska Knjiga Publishing House, Beograd, (1966), Vol. **1**, pp. 464–467. G. Lisitano; "Radio-frequency Plasma Generator", U.S. Patent 3,663,858, issued in 16 May 1972, priority 6 November 1970.

[4.6] G. Lisitano, R.A. Ellis, Jr., W.M. Hooke, and T.H. Stix: "Production of quiescent discharge with high electron temperatures"; Rev. Sci. Instrum. **39** (3) (1968) 295–297.

[4.7] R.M. Fredericks and J. Asmussen: "High-density resonantly sustained plasma in a variable-length cylindrical cavity"; Appl. Phys. Lett. **19** (1971) 508–510.

[4.8] J. Musil: "Deposition of thin films using microwave plasmas: present status and trends"; Vacuum **47** (2) (1996) 145–155.

[4.9] Y.A. Lebedev: "Microwave discharges at low pressures and peculiarities of the processes in strongly non-uniform plasma"; Plasma Sources Sci. Technol. **24** (2015) 053001 (39pp).

[4.10] L. Bardos, G. Loncar, I. Stoll, J. Musil, and F. Zacek:: "A method of formation of thin oxide films on silicon in a microwave magnetoactive oxygen plasma"; J. Phys. D: Appl. Phys. **8** (1975) L195–L197.

[4.11] R. Dragila, L. Bardos, and G. Loncar: "Growth of SiO2 films on Si in an oxygen microwave discharge"; Thin Solid Films **34** (1976) 115–117.

[4.12] J. Musil, F. Zacek, L. Bardos, and J. Datlov: "Efficient microwave source of a dense magnetoplasma"; Czech. J. Phys. **B28** (1978) 74–76.

[4.13] L. Bardos, J. Musil, F. Zacek, and L. Hulenyi: "The negative role of fast electrons in the microwave oxidation of silicon"; Czech. J. Phys. **B28** (1978) 639–643.

[4.14] J. Musil, F. Zacek, L. Bardos, G. Loncar, and R. Dragila: "Plasma oxidation of silicon in a microwave discharge and its specificity"; J. Phys. D: Appl. Phys. **12** (1979) L61–L63.

[4.15] G. Loncar, J. Musil, and L. Bardos: "Present status of thin oxide films creation in a microwave plasma"; Czech. J. Phys. B **30** (1980) 688–707.

[4.16] L. Bardos, R. Dragila, G. Loncar, and J. Musil: "Creation of thin oxide films in a microwave oxygen plasma"; Academia (Praha 1983). ISSN0069-228X. https://katalog.muni.cz/Record/MUB01000347222/Details

[4.17] S. Altmannshofer, J. Boudaden, R. Wieland, I. Eisele, and C. Kutter: "Microwave plasma assisted process for cleaning and deposition in future semiconductor technology"; IOP Conf. Series **213** (2017) 012021. http://dx.doi.org/10.1088/1757-899X/213/1/012021.

[4.18] S. Fujimura, T. Kisa, and Y. Motoki: "Microwave plasma processing process and apparatus", US Patent 5,364,519, 15 November 1994.

[4.19] X. Liu, J. Hao, N. You, Y. Bai, and S. Wang: "High-pressure microwave plasma oxidation of 4H-SiC with low interface trap density"; AIP Adv. **9** (2019) 125150. http://dx.doi.org/10.1063/1.5115538.

[4.20] H. Xu and R.L. Bersin: "Cleaning and stripping of photoresist from surfaces of semiconductor wafers", US Patent 5,908,319, 1 June 1999.

[4.21] S. Pal, R. Mahapatra, S.K. Ray, B.R. Chakraborty, S.M. Shivaprasad, S.K. Lahiri, and D.N. Bose: “Microwave plasma oxidation of gallium nitride””; Thin Solid Films **425** (2003) 20–23.

[4.22] H.S. Oon and K.Y. Cheong: “Recent development of gallium oxide thin film on GaN”; Mater. Sci. Semicond. Process. **16** (2013) 1217–1231.

[4.23] L.G. Christophorou, R.N. Compton, G.S. Hurst, and P.W. Reinhardt: “Determination of electron-capture cross sections with swarm-beam techniques”; J. Chem. Phys. **43** (12) (1965) 4273–4281.

[4.24] J.B. Hasted: “Physics of atomic collisions”, Butterworth, London, 1964.

[4.25] T. Ishikawa, D. Hayashi, K. Sasaki, and K. Kadota: “Determination of negative ion density with optical emission spectroscopy in oxygen afterglow plasmas”; Appl. Phys. Lett. **72** (19) (1998) 2391–2393.

[4.26] V. Kopecky, J. Musil, and F. Zacek: “Absorption of microwave energy in a plasma column at high magnetic fields”; Phys. Lett. **50A** (4) (1974) 309–310.

[4.27] A.E. Aubert, A.M. Messiaen, and P.E. Vandenplas: "Explanation of Plasma Densities High Above Critical Density in a Cylinder Excited by an Rf Helix"; Appl. Phys. Lett. **18** (2) (1971) 63–65.

[4.28] J. Musil and F. Zacek: “Experimental study of the absorption of intense electromagnetic waves in a magnetoactive plasma”; Rozpravy CSAV, Academia (Praha 1975).https://rusneb.ru/catalog/000200_000018_RU_INFOCOMM_322_NLR_5000210840/

[4.29] V.M. Shibikov:: “Fast electron generation in a pulsed discharge plasma in helium at a pressure of 0.02–2.0 Torr”; Plasma Phys. Rep. **46** (2) (2020) 230–235. ISSN 1063-780X.

[4.30] L. Bardos and J. Musil:: “Axial decaying of the microwave ECR oxygen plasma”; J. Phys. D: Appl. Phys. **21** (1988) 1459–1461.

[4.31] L. Bardos: “Afterglow and decaying plasma CVD systems”; Vacuum **38** (8–10) (1988) 637–642.

[4.32] Y. Zhang, C. Charles, and R.W. Boswell:: “Density measurements in low pressure, weakly magnetized, RF plasmas: experimental verification of the sheath expansion effect”; Front. Phys. **5** (2017) article 27, 1–6.

[4.33] J. Musil and F. Zacek: “Localization of a high-power microwaves absorption in the cylindrical plasma column at $\omega_{LH} \ll \omega \ll \omega_{ce}$”; Czech. J. Phys. **B28** (1978) 533–535.

[4.34] S.C. Brown: “Basic data of plasma physics”, Wiley, New York, 1960.

[4.35] T. Iyasu and R. Shimizu: “Database construction of secondary electron emission – Monte Carlo approach combined with supplementary experiment”; J. Surf. Analys. **14** (4) (2008) 312–315.

[4.36] S.J. Jokela, I.V. Veryovkin, A.V. Zinovev, et al and the Large Area Picosecond Photodetector Collaboration: “Secondary electron yield of emissive materials for large-area micro-channel plate detectors: Surface composition and film thickness dependencies"; Phys. Procedia **37** (2012) 740–747.

[4.37] M. Koitabashi, M. Yananoto, H. Kimura, A. Hattori, T. Watanabe, and T. Nakano: "Secondary electron emission characteristics of heavy-metal oxide coatings for CRT shadow mask"; Jpn. J. Appl. Phys. **27** (7) (1988) 1210–1215.

[4.38] L. Bardos, J. Musil, and P. Taras:: "Differences between microwave and RF activation of nitrogen for the PE CVD process"; J. Phys. D: Appl. Phys. **15** (1982) L79–L82.

[4.39] L. Bardos, J. Musil, and P. Taras: "Nitrogen activation for plasma chemical synthesis of thin $Si_3N_4$ films"; Thin Solid Films **102** (1983) 107–110.

[4.40] L. Bardos, J. Musil, and M. Lubanski: "Chemiluminescence of the silane – active nitrogen reactions during PE CVD of the silicon nitride films"; Czech. J. Phys. **B34** (1984) 1242–1245.

[4.41] M. Moisan, C. Beaudry, and P. Leprince: "A new HF device for the production of long plasma columns at a high electron density"; Phys. Lett. **50A** (1974) 125–126.

[4.42] M. Moisan, Z. Zakrzewski, and R. Pantel:: "The theory and characteristics of an efficient surface wave launcher (surfatron) producing long plasma columns"; J. Phys. D: Appl. Phys. **12** (1979) 219–237.

[4.43] M. Moisan and Z. Zakrzewski: "Plasma sources based on the propagation of electromagnetic surface waves"; J. Phys. D: Appl. Phys. **24** (1991) 1025–1048.

[4.44] R.W.B. Pearse and A.G. Gaydon: "The identification of molecular spectra", 3rd Edition, J. Wiley and Sons, Inc., New York, 1963.

[4.45] L. Bardos and J. Musil: "Silicon nitride films prepared by PA CVD outside the plasma"; Czech. J. Phys. **B35** (1985) 1437–1444.

[4.46] L. Bardos and H. Baránková: "Processing in afterglows and decaying plasmas", Vacuum Technology & Coating, Weston, CT, Septembar 2001, pp. 46–56.

[4.47] C. Blaauw: "Preparation and characterization of plasma-deposited silicon nitride"; J. Electrochem. Soc. **131** (5) (1984) 1114–1118.

[4.48] S.W. Lim, Y. Shimogaki, Y. Nakano, et al.: "Decrease in deposition rate and improvement of step coverage by $CF_4$ addition to plasma-enhanced chemical vapor deposition of silicon oxide films"; Jpn. J. Appl. Phys. **19** (2000) 330–336.

[4.49] L. Bardos and S. Berg: "Microwave apparatus for plasma processing"; Swedish Patent 9302222-6, June 1993.

[4.50] L. Bardos, H. Baránková, and S. Berg:: "Microwave surfatron system for plasma processing"; J. Vac. Sci. Technol. **A14** (2) (1996) 474–477.

[4.51] L. Bardos, H. Baránková, Y.A. Lebedev, et al.: "Diamond deposition in a microwave electrode discharge at reduced pressures"; Diamond Rel. Mat. **6** (1997) 224–229.

[4.52] L. Bardos, H. Baránková, and Y.A. Lebedev: "Effective low power microwave plasma CVD of carbon nitride films"; Proc. of 42nd Technical Conference of Society of Vacuum Coaters, April 17–22, 1999, Chicago, Paper E-7, p. 111–115.

[4.53] L. Bardos, H. Baránková, Th, Welzel, et al.: "Comparison of the radio frequency hollow cathode to the microwave antenna discharge for plasma processing"; J. Appl. Phys. **90** (4) (2001) 1703–1709.

[4.54] L. Bardos, H. Baránková and A. Bardos: "High Rate Low Power Microwave Plasma CVD of Carbon Nitride Films"; Presented at the 1999 Joint International Meeting of The Electrochemical Society and The Electrochemical Society of Japan, Hawaii, October 17-22 1999. Paper no. 791. Abstract book vol. 99-2.

[4.55] R. Pintaske, Th. Welzel, M. Scheller, et al.: "Spectroscopic studies of a magnetron sputtering discharge for boron nitride deposition"; Surf. Coat. Technol. **99** (3) (1998) 266–273.

[4.56] L. Bardos and Y.A. Lebedev: "Features of a nonequilibrium microwave electrode discharge"; Plasma Phys. Rep. **24** (10) (1998) 891–895.

[4.57] L. Bardos and H. Baránková: "Device for hybrid plasma processing"; Swedish patent No. 9904295-4, priority 1999-11-26, WO0139560 (A1) – 2001- 05-31.US6899054 (B1), JP2003515433 (A).

[4.58] L. Bardos and H. Baránková: "An apparatus for generation of a linear arc discharge for plasma processing"; Swedish patent No. 503 141, U.S. Pat. 5,908,602 priority November 1994.

[4.59] H. Baránková and L. Bardos: "Plasma processing apparatus having rotating magnets"; U.S. patent 6,351,075, 20 November 1997.

[4.60] L. Bardos, H. Baránková, and S. Berg:: "Linear arc discharge source for large area plasma processing"; Appl. Phys. Lett. **70** (1997) 577–579.

[4.61] H. Baránková and L. Bardos: "Hollow cathode plasma sources for large area surface treatment"; Surf. Coat. Technol. **146–147** (2001) 486–490.

[4 .62] L. Bardos, H. Baránková, L.-E. Gustavsson, and D.G. Teer: "New microwave and hollow cathode hybrid plasma sources"; Surf. Coat. Technol. **177–178** (2004) 651–656.

[4.63] J.A. Thornton:: "The microstructure of sputter-deposited coatings"; J. Vac. Sci. Technol. A **4** (1986) 3059–3065.

[4.64] L. Bardos, L.-E. Gustavsson, and H. Baránková: "Effect of ferromagnetic substrates on the film growth in magnetized plasma systems"; Surf. Coat. Technol. **200** (2005) 1862–1866.

[4.65] L.-E. Gustavsson, H. Baránková, and L. Bardos: "Some properties of TiN films produced in hollow cathode and microwave ECR hybrid plasma system"; Surf. Coat. Technol. **201** (2006) 1464–1468.

# 5

# Microwave Plasma Systems at Atmospheric and Higher Pressures

In Section 2.2, Chapter 2, we introduced the Paschen curve, which showed the function $V_B = f(p\ d)$ and described how the breakdown voltage ($V_B$) depends on the gas pressure p and the distance d between two electrodes. The Paschen curve has a typical valley shape, which is similar for all gases; however, the values of p, d and $V_B$ are different. Figure 5.1 shows a detailed Paschen curve for air.

In contrast to the low-pressure breakdown, the necessary breakdown voltage in dry air at atmospheric pressure in a gap of 1 cm between electrodes is about 30 kV, even when the distance is $d \approx 1$ mm the ignition of the plasma still requires 3.2 kV in air, 1.5 kV in Ar and 0.75 kV in He, see Refs. [5.1, 5.2]. This explains why the ignition of atmospheric plasma needs a high voltage or small distance between electrodes. Due to high collision frequency and corresponding short mean-free paths, the atmospheric plasma usually has a limited spatial outreach. It will be shown later, e.g. in Section 5.2, Figure 5.8, that the outreach of the plasma can be prolonged by the flowing gas or by working in some kind of a decaying plasma or in a near afterglow. These terms were explained in Section 3.4, Chapter 3.

If the breakdown is performed by a radio frequency (RF) power (the frequency order of 10 MHz) instead of a DC power, the amplitude of the breakdown electric field is somewhat lower. Our measurements of the RF breakdown at atmospheric pressure in fused hollow cathode systems with about 1 mm gap between the RF biased cathode and a grounded counter electrode showed the peak-to-peak voltages of 1.2 kV in air, 0.4 kV in Ar and 0.18 kV in Ne, see Ref. [5.3]. However, with a frequency increase, the breakdown voltage can increase again. Measurements, reported in Ref. [5.4], performed at 2.86 GHz with 180 ns pulses in air at atmospheric pressure between two opposite pins in waveguide, showed about 6.2 kV/mm in air. On the other hand, the study of the gas breakdown in a discharge tube passing through a 2.4 GHz rectangular $TE_{10}$ waveguide cavity section (similar to the arrangement in Figure 2.11a, Section 2.5.3, Chapter 2), showed lower values, e.g. only about 0.14–0.17 kV/mm for Xe, Kr and Ar, see Ref. [5.5]. In Ref. [5.6], the breakdown in air at 9.4 GHz was estimated to 0.28 kV/mm. The interpretation of the

*Microwave Plasma Sources and Methods in Processing Technology*, First Edition.
Ladislav Bárdoš and Hana Baránková.

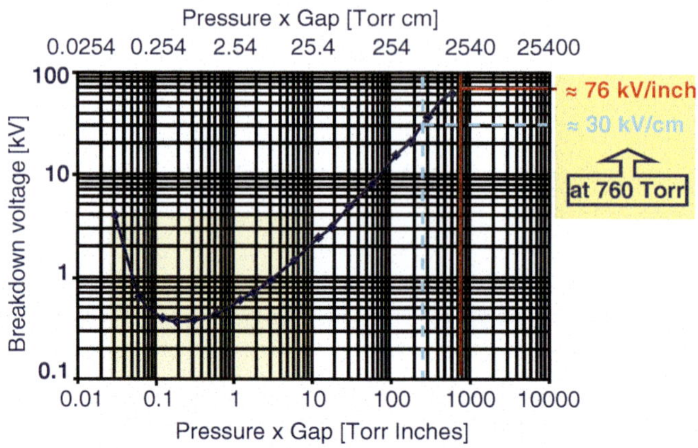


**Figure 5.1** The Paschen curve data for breakdown voltage ($V_B$) measured for dry air between DC-biased electrodes at the distance gap of one inch (25.4 mm). The highlighted area in the graph represents typical parameters used in reduced-pressure and low-pressure plasma processing devices.

breakdown measurements is always complicated because the breakdown is affected by the secondary electron emission coefficient of the metal in used electrodes and by electrode shapes. Because the atmospheric breakdown voltage is often considerably higher than the voltage needed for maintaining the discharge, many atmospheric plasma sources, particularly those working with air and molecular gases, require certain ignition tools, e.g. sparks, high voltage spikes.

## 5.1 Features of the Atmospheric Plasma and Cold Atmospheric Plasma (CAP) Sources

Contrary to the well-known and established thermal plasmas (used mostly in pyrolytic processes, e.g. in metallurgy, waste incinerations, or spray coating technology, at high powers, and at energy equilibrium between particles), the cold plasmas at the energy non-equilibrium use rather low-power generators with special electrode-launching or power-launching- arrangements. Many sources and processes have been described and characterized in publications, e.g. see review in Ref. [5.7]. Moreover, a few reports and extensive reviews describe different principles and applications of the CAP, with numerous related references, which reflects an enormous interest in this technology. For example, recent publications describe the following:

- Interactions of cold plasma with solid surfaces for surface activation and improved adhesion of subsequent coatings or paints, see Refs. [5.8–5.10].

- The PE CVD of silica films, see Ref. [5.11].
- Processes of impregnation and treatment of textiles, see Ref. [5.12].
- Deposition of carbon nanotubes, see Ref. [5.13].
- Interactions of atmospheric plasma with liquids, see Refs. [5.14, 5.15].
- Treatments of implants or biological and organic surfaces including food, see Refs. [5.16–5.19] and plasma medicine, see Refs. [5.20–5.23].
- Applications in the agriculture for enhanced germination of seeds, see Refs. [5.24–5.26].

More examples and new trends in the use of cold atmospheric plasma (CAP) will be described also in Chapter 6.

At one atmosphere and room temperature, the mean-free paths are roughly between 0.1 and 1 μm, and the high collision frequency has the tendency to quickly settle the energy equilibrium between plasma particles. The high-energy heavy particles (molecules, atoms, ions) at high pressure composes typically a thermal (hot, or equilibrium) plasma. On the other hand, the generation of cold plasmas (low energy, non-thermal, non-equilibrium) at high pressures requires *selective delivery* of the power from the generator to the electrons and not to heavy particles. Therefore, arranging the plasma generation system is necessary to prevent the heavy particles from absorbing the delivered power. Several methods enable such requirements. The five most typical ones are characterized below.

(1) ***High-frequency AC electric fields and short DC pulses***:
The high-frequency field is an advantage for the microwave power. It leads to the decoupled movements of electrons and ions due to their different masses, i.e. we know that the electrons are about 1000x lighter than ions. In the high-frequency fields, the space-charge sheaths are thinner, ion energies are lower, and the electrode phenomena become less important. Moreover, the high-frequency AC and pulsed DC discharges can be generated without contact of the electrodes with the plasma gas, and they can work as electrodeless glow plasma. Due to a short incidence time, the nanosecond DC pulses (shorter than about 100 ns) cannot form microarcs (streamers) and filaments. Such protection against arcing does not need any help from some of the dielectric barriers, see Ref. [5.27].

(2) ***Selection of suitable gas***:
Molecular gases are generally more complicated than atomic ones because they tend to absorb power into their inner energy states (e.g. to the vibration or rotation of the atoms in the molecules). Therefore, among the most complicated molecular gases are air, $O_2$, and $CO_2$. The atomic gas is easier to work with. The best choice for the atmospheric plasma is helium (He) because of several reasons. The helium plasma allows the fastest production of secondary electrons by the gamma processes (see Townsend coefficient gamma in Section

2.2, Chapter 2) because the He atoms are small (the He atom has diameter 2.8 Å, air molecule has 9.7 Å) and exhibit lower collision frequency at atmospheric pressure than other gases (about 10 GHz in He compared to about 60 GHz in air). Moreover, a high thermal conductivity in helium gas can suppress formation of local hot streamers.

**(3) *Fast-flowing gas and stabilization of the plasma by high gas flow rates*:**
The high rates of the flowing gas have obvious cooling effects, which prevent transitions of the glow discharge into the hot arc. Fast-flowing gas particles can be expected to be in the plasma generation area for a shorter time (i.e. residence time). Moreover, the fast-flowing gas is beneficial for the expanding of the plasma and increasing its spatial outreach. It is often used in different microwave powered atmospheric plasma jets at powers of several kilowatts.

**(4) *Dielectric barriers between electrodes = suppressed corona = DBD*:**
Dielectric barriers prevent the ohmic currents and limit the current values. Material of the dielectric barrier discharge (DBD) is important because of different charging and an assistance in the Penning ionization. Electrets are useful, e.g. Mylar. In the DBD systems, only AC or pulsed DC fields can generate plasma because the conductive (ohmic) DC currents cannot flow through the dielectric and the power circuit allows only displacement AC currents (like in the capacitors). An accumulated charge on the dielectric during one polarity of the electric field is compensated by an opposite charge during an opposite polarity of the alternating electric field.

**(5) *Special shapes of the electrodes*:**
Sharp electrode edges create high electric fields, which help in the electric breakdown. Brush-type electrodes, see Ref. [5.28], and mesh-type electrodes, see Ref. [5.29], were proved to keep the glow-type discharges in the DBD systems without filaments (streamers) representing the local filamentary arcs.

Probably the first CAP system was presented in 1857 by W. Siemens, see Ref. [5.30], and in 1860 by T. Andrews and P.G. Tait, see Ref. [5.31]. Both these systems were based on the DBD arrangement. Siemens used his device for the production of ozone. In 1941, G.D. Cristescu and R. Grigorovici reported a high-frequency (10–70 MHz) coaxial electrode arrangement, see Ref. [5.32], followed by L. Mollwo in 1958 with a microwave coaxial (940 MHz) system capable to generate about 2 cm long torch in air, see Ref. [5.33]. Figure 5.2 shows the Mollwo's microwave plasma torches.

Figure 5.3 shows schematic illustrations of the most common CAP devices generated by an AC, pulsed DC, or RF power with different arrangements of dielectric barriers.

Figure 5.4 shows microwave and very-high-frequency powered CAP sources. See also Refs. [5.2, 5.7, 5.34, 5.35].

**Figure 5.2** Plasma torches in air at atmospheric pressure at 32 cm wavelength and about 100 W power, see Ref. [5.33]. The longer torch of about 2 cm long had an estimated temperature of 3500 °C.

Most of the microwave power systems for generation of atmospheric plasma is based on a waveguide power lines with simple quartz reactor tube, like in Figure 5.4a. However, arrangements of the microwave launchers for generation of the atmospheric plasma can be designed with different shapes and sizes of electrodes. Interestingly, even simple microwave ovens (Section 1.5,

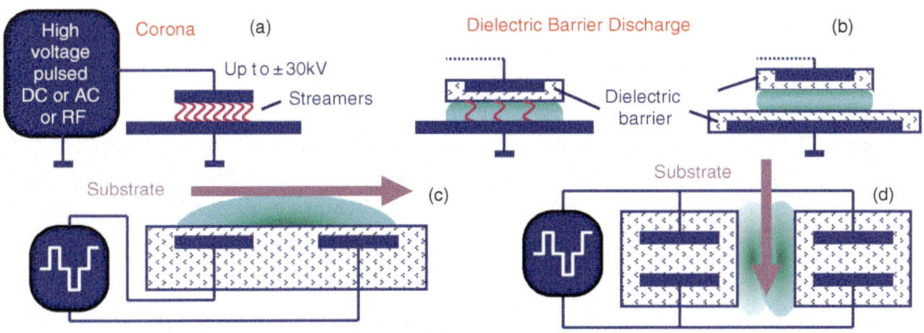


**Figure 5.3** Examples of the most frequently used cold atmospheric plasma systems and their arrangements for treatment of substrates (or gases). (a) Simple corona discharge. (b) Dielectric barrier DBD systems with one or both electrodes screened by the dielectric. (c) Coplanar DBD system for large area applications, Based on Ref. [5.34]. (d) A DBD for processing at both sides.

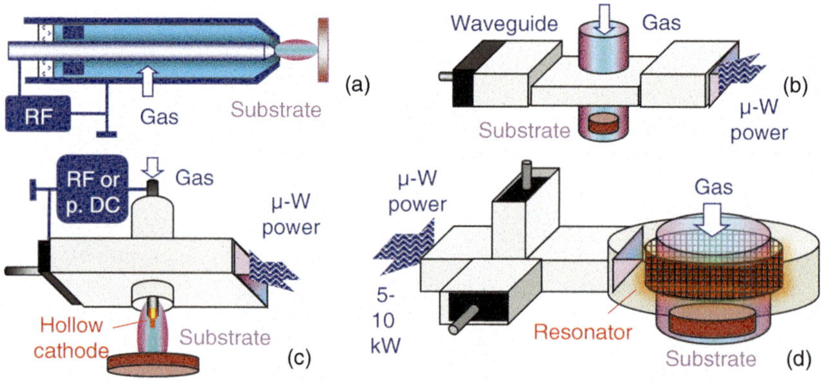


**Figure 5.4** Examples of the high frequency and microwave powered cold atmospheric plasma systems and arrangements. (a) Simple coaxial discharge system by Mollwo, Based on Ref. [5.33]. (b) Rectangular waveguide system with the dielectric reactor passing the tapered section. (c) Novel hybrid system with a microwave antenna working as a hollow cathode, Based on Ref. [5.37]. (d) High microwave power system with special resonator for large diameter reactors (e.g. see Ref. [5.38]).

Chapter 1) are used for atmospheric plasma experiments and surface processing applications, e.g. see Ref. [5.39]. The microwave oven can be used for underwater plasma processes with the special shape of submerged electrode focusing the microwave power and generating the plasma, see Ref. [5.40]. The microwave atmospheric plasma in liquids will be described in more details in Section 5.4.

## 5.2 Atmospheric Microwave Plasma Sources Assisted by Hollow Cathodes

The microwave system shown in Figure 5.4c has a non-conventional design, and its construction and applications will, therefore, be described in more detail. This novel hybrid source with a microwave antenna, which can work simultaneously as a hollow cathode, was patented in 1999, see Ref. [5.41]. The features and applications of this source were presented in several papers, see Refs. [5.37, 5.42–5.45]. The source is the Hybrid Hollow Electrode Activated Discharge (H-HEAD), and its principle is illustrated schematically in Figure 5.5.

The principle of the H-HEAD source, shown schematically in Figure 5.5, is similar to the Surfajet device working at reduced gas pressures and described in Figure 4.35 (Section 4.3, Chapter 4) and in Figure 4.47a (Section 4.3.2, Chapter 4). It has a coaxial arrangement installed in a waveguide, similar to the waveguide Surfatrons, see Ref. [5.46], but it is arranged with a central metallic

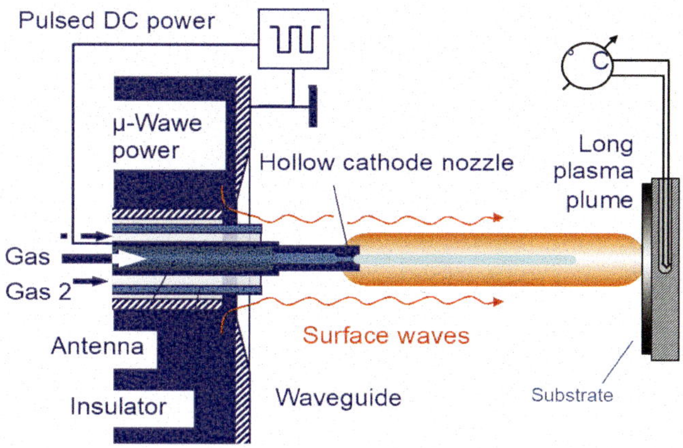


**Figure 5.5** The H-HEAD source with a microwave antenna working as a hollow cathode.

antenna terminated by a gas nozzle, which can serve as a hollow cathode powered by DC, RF, or a pulsed DC power.

The microwave power forms surface waves propagating along the antenna and further along the interface between the generated long plasma column (i.e. the plasma plume) and the surrounding air. The microwave surface-wave-generated plasma can help in starting the hollow cathode plasma and vice-versa the hollow cathode plasma can help in starting the microwave surface-wave plasma. In this hybrid arrangement, the microwave plasma can control the plasma density and the hollow cathode can control energy of the charged particles. The system allows combinations of gases if required for certain plasma-chemical reactions. Figure 5.6 shows the effect of the hollow cathode on the performance of the argon plasma.

The hollow cathode effect inside the 0.5 mm diameter nozzle is connected with an enhanced ionization due to oscillating electrons between inner walls of the nozzle and an ion bombardment of the inner walls, which results in an evident heating of the nozzle walls. Figure 5.6 shows the contribution of the hollow cathode was detected on an increasing temperature of the substrate holder positioned at 25 mm distance from the source.

The substrate temperature at the microwave power of 400 W is relatively high, see Figure 5.6. However, at power $\leq$ 50 W, the plasma plume is not hot, see Figure 5.7.

The authors's previous experience with different low-power microwave generators for the CAPs confirmed somewhat better performance of classical vacuum-tube-driven high-voltage electronics compared to the advanced semiconductor-based microwave generators. The tube-driven high-voltage supply for the magnetron generator gave more stable power from about 10 W.

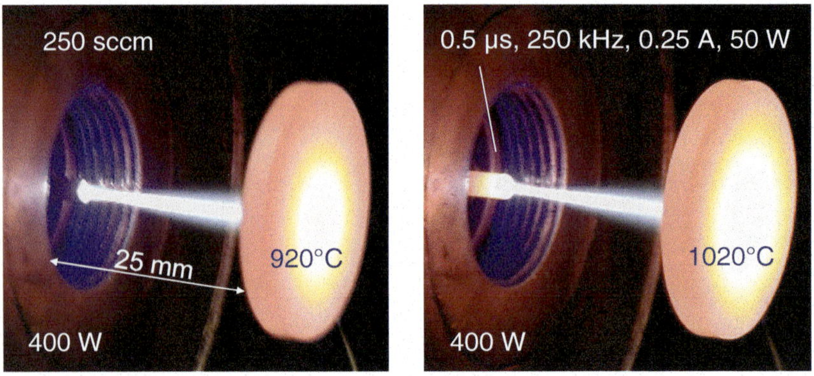


**Figure 5.6** The effect of the pulsed DC hollow cathode (0.5 μs, 250 kHz, 50 W) on the power transferred by the plasma plume in the H-HEAD plasma source. The nozzle became hot.

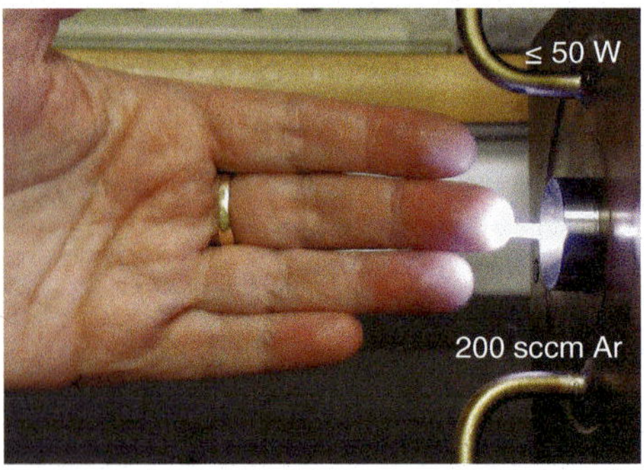


**Figure 5.7** The cold plasma jet at low power generated without a hollow cathode.

The more modern semiconductor electronics was more rigid and remained stable rather for higher powers exceeding 100 W. However, this might not be the case with the newest generators.

Due to specific surface-wave generation, the H-HEAD source could form about 200 mm long plasma plumes in argon and about 60 mm long plumes in air, see Figure 5.8.

As it can be seen in Figure 5.8, in addition to the possibility to generate long atmospheric plasma plumes, the source can work at low gas flows compared to the flows used in the low-pressure plasma systems. This represents a non-conventional and unique feature of the H-HEAD sources. The reason for such an advantage is given by the dependency of the length of the surface-wave transport on the interface between the plasma and the surrounding air. If the gas flow is laminar, this interface is smooth, and the surface waves can propagate to long distances and produce long plasma plumes. By increasing the gas flow rate, the interface forms eddies connected with the turbulent flow.

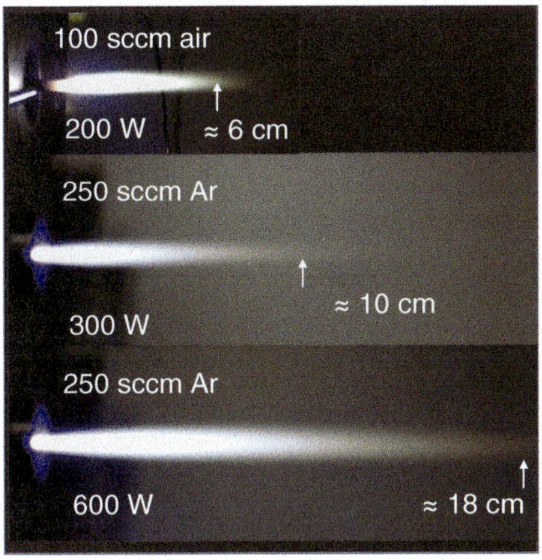


**Figure 5.8** Comparison of lengths of the atmospheric plasma jets generated in air and in argon by the H-HEAD source without the contribution of the hollow cathode, see also Ref. [5.43].

The interface is not smooth any more, which shortens the propagation distance of the microwave surface waves and, therefore, the length of the plasma column. Figure 5.9 shows an experimental proof of these effects in argon plasma plumes generated at 400 W microwave power.

The principle of the hybrid atmospheric plasma source was tested on a novel construction of a brush plasma so as to test the feasibility of an atmospheric plasma system for large-area processing. In the laboratory conditions with

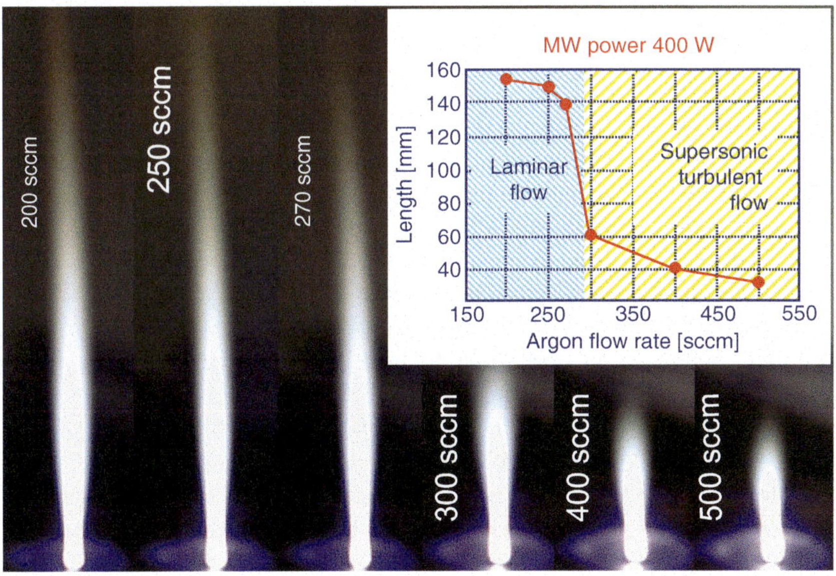


**Figure 5.9** Dependency of the argon plasma plume length at 400 W power in the H-HEAD source on the argon flow rate.

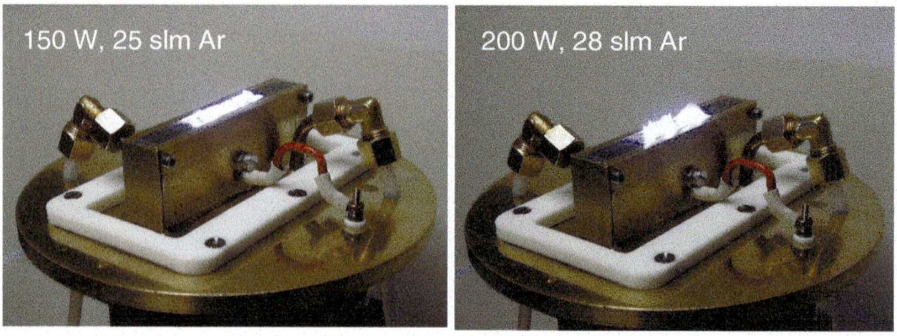


**Figure 5.10** The test equipment for generation of a brush-shaped atmospheric plasma (unpublished results).

limited available gas flow rates, the system was generated at Ar flow rate up to 30 slm (30 000 sccm), see Figure 5.10.

The microwave antenna in this construction was represented by several pairs of parallel plates, and the stable plasma regime depended on the higher gas flow rates. The limited gas flow rate limited the maximum applicable power due to heating of the antenna plates. The possibility of broader plasma areas was confirmed, but at used conditions, the plasma had limited spatial outreach.

### 5.2.1 Applications of the H-HEAD Plasma Source in Surface Treatments

An advantage of the H-HEAD sources is a long plasma plume though one disadvantage is higher microwave power (several hundreds of watts) needed to sustain the stable plasma for processing applications. In most applications, at large surface areas, the plasma plume must be multiplied to several plumes above the treated surface or arranged with a robotic arm system for multiple scans of the plume above the surface. The H-HEAD source was successfully tested for surface activations of several materials, including heat sensitive polymers, where the surface temperature was controlled by longer distances of the surface from the plasma source and by the scanning rate of the plasma plume over the surface. Figure 5.11 shows an example of the measured *contact angle* after surface activation of the steel vs. number of scans. The contact angle is measure of the surface vettability (lower angle is better vettability).

An increased surface energy is also measured by a decrease in the contact angle of a water drop, as shown in Figure 5.11 on steel surface. The activated surfaces are more hydrophilic. The H-HEAD plasma treatment can be applied in a similar way like a lacquer spraying pistol. For certain materials, the activated surface can last a long time. The surface activation on a bulk carbon-based substrate increased the surface energy from 41 mN/m to 71 mN/m in a few minutes, see Ref. [5.45]. The surface activity can then last for more than a

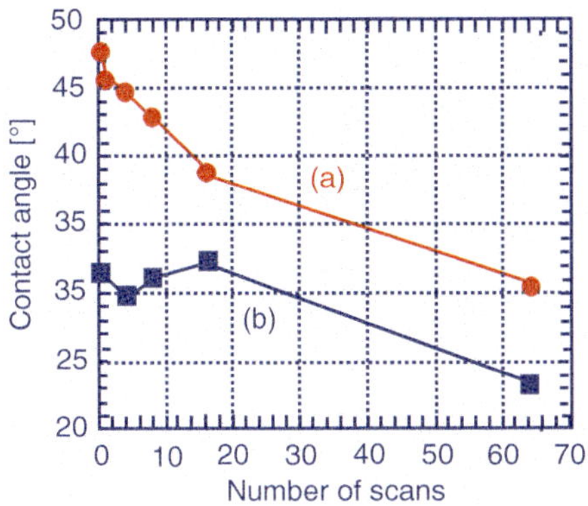


**Figure 5.11** The contact angles on the H-HEAD plasma treated DC 01 steel as a function of the number of scans of the plasma plume on the surface. (a) The data was acquired 8 s after triggering the pendant drop. (b) The data after 3 min stabilization of the sessile water drop. Based on Ref. [5.45].

half year. Figure 5.12 shows an example of the carbon fiber material used in construction of Airbus planes and treated by the H-HEAD plasma on the half surface. The surface energy was increased from 40 to 70 mN/m and lasted for more than one year, see Ref. [5.47]. Because the activated surface is hydrophilic, it is suitable for pre-treatments before final lacquering or gluing with other materials. Therefore, surface activation can replace primer lacquer before final lacquering.

Contrary to the plastics and polymer materials, the surface activity of metals lasts for a shorter time, in the most cases less than one hour. Therefore, any subsequent treatments should be made within shorter times (about 10–20 minutes). The surface activation documented in Figure 5.11 was made on ordinary construction steel DC 01 used, e.g. in the construction of buoys for ocean wave power converters.

Because the lifetime of ocean buoys depends on the anticorrosion protection of their surfaces, it depends on the adhesion of the protective lacquers. The adhesion of the lacquer on the steel surface (without using any primer) improved after a short treatment by the H-HEAD plasma, see Figure 5.13.

A static interaction of the H-HEAD plasma in argon, see Figure 5.6, with the silicon sample was able to form 0.5 μm thick $SiO_2$ film during several

**Figure 5.12** An increased hydrophobicity on the right part of the carbon fiber sample after several scans by the H-HEAD plasma.

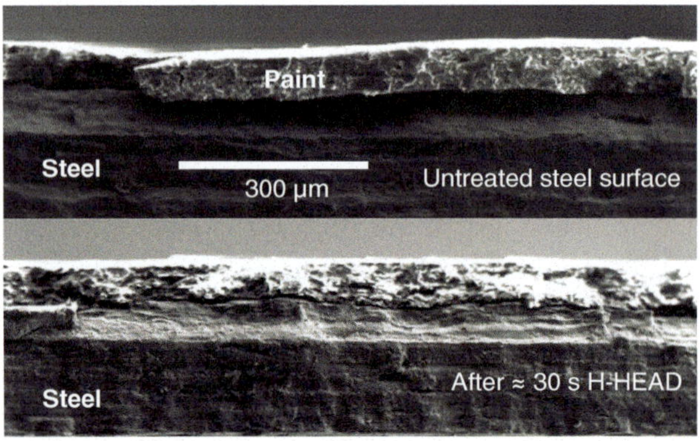


**Figure 5.13** Improvement of the lacquer adhesion on DC 01 steel after H-HEAD treatment. Parameters: 2.4 GHz, 240 W, 300 sccm air passing the water bubbler at room temperature.

seconds, see Figure 5.14. This was faster than in low-pressure oxidations, e.g. at oxygen pressure $p_{O2}$ = 0.27 Pa (2 mTorr), described in Chapter 4 (see Figures 4.14 and 4.15, Section 4.1.2).

Even at 2.5 cm distance between the substrate and the H-HEAD source traces of oxides of the metal from the hollow cathode nozzle were found on the substrate surface. The Ti nozzle caused measurable contents of $TiO_2$ in the $SiO_2$ spot, see Ref. [5.37]. Moreover, the components of the stainless steel

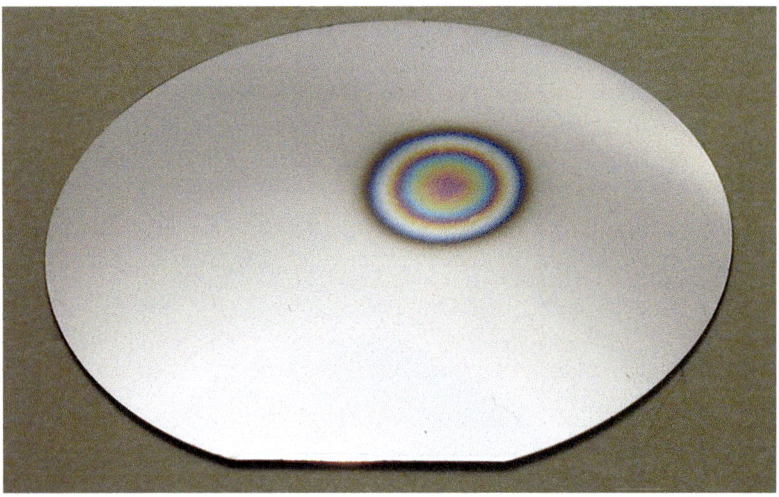

**Figure 5.14** About a 5-second oxidation of the Si sample by H-HEAD Ar plasma in open air. Parameters: 250 sccm Ar, 2.4 GHz, 400 W, hollow cathode pulsed DC 50 W, 0.5 μs, 250 kHz, 0.25 A. Sample distance from the H-HEAD nozzle was 2.5 cm.

nozzle were detected by optical spectroscopy in the plasma. An example of the Fe optical emission (line 381.9 nm) detected at 5 mm distance from the steel nozzle is shown as a function of the hollow cathode pulsed current in Figure 5.15.

Therefore, the contribution of the hollow cathode can be beneficial when the plasma should have higher density and ion energy. However, the hollow cathode nozzle can also produce metal particles. To reduce the presence of metal particles from the hollow cathode on the substrate, the H-HEAD system should be arranged at rather long distances from the substrate, at least about 5 cm. To avoid all metal particles, the system should be operated only with the microwave antenna without hollow cathode at powers under 200 W.

The contribution of the hollow cathode plasma in the H-HEAD device was beneficial in several processes requiring higher temperatures. An example of such applications is the plasma sintering of powders. Figure 5.16 shows the plasma sintering of powder. Sintering tests on special diamond-coated powders confirmed the possibility producing hard compact solid materials in several minutes. Such composite materials can be useful in special hard tools, e.g. for grinding applications.

The H-HEAD source could be tuned by the microwave power and the pulsed DC power for sensitive control of the sintering temperature and, consequently, for the best properties of the resulting compact material. Additional control parameters are the gas flow rate, selected gas (mostly argon), and the distance between the nozzle of the H-HEAD device and the holder with the powder.

A surprising result was found in the PE CVD by the air plasma plume with an alcohol ($C_2H_5OH$) vapor carried by air. The result was a film composed of nanocluster diamonds. Such films were synthesized on the unpretreated substrates, e.g. on molybdenum or on stainless steel. At the microwave power of

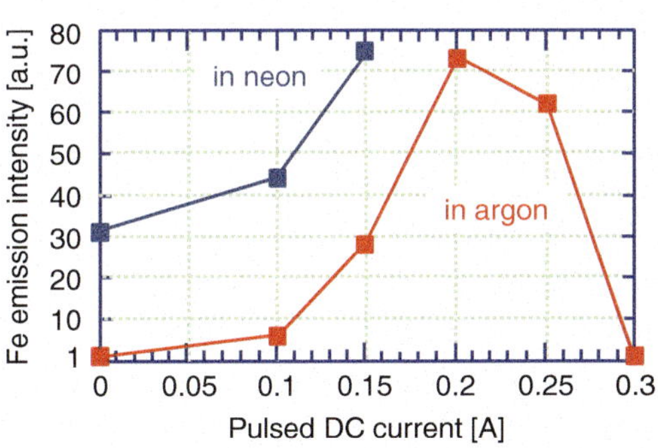


**Figure 5.15** Dependence of the optical emission intensity of the Fe line 3819 Å in neon and argon plasma vs. pulsed DC current in the Fe hollow cathode circuit. Parameters: optical selector at z = 5 mm; grounded anode (sample holder) 20 mm from the cathode; microwave power 500 W; gas flow rate 200 sccm.

**Figure 5.16** The H-HEAD plasma sintering of powders. Parameters on the photograph: 250 sccm Ar, 2.4 GHz, 400 W, hollow cathode pulsed DC 50 W, 0.5 μs, 250 kHz, 0.25 A. The substrate holder was 30 mm in diameter. The holder was fixed on 1.5 mm diameter stainless steel cartridge of the K-type thermocouple measuring the temperature of the substrate holder.

**Figure 5.17** Experimental arrangement with the H-HEAD source generating the air plasma for deposition of nanocrystalline diamond clusters. The substrate is 0.25 mm thick stainless steel.

**Figure 5.18** The scanning electron microscope (SEM) image of nanocluster grains in the diamond coating after a one-hour PE CVD in the H-HEAD air plasma interacting with the mixture of air with alcohol.

around 300 W and the gas flow rates below 300 sccm, the temperature caused by the air plasma plume on a thermally insulated substrate placed 1.5 cm from the hollow cathode was less than 600°C. At these conditions the 200 nm thick coatings with nanocluster diamond grains and aggregates were formed on the unpretreated substrates after one hour, see Ref. [5.44]. Figure 5.17 shows the used experimental arrangement, and Figure 5.18 shows the resulting nanoclusters. The characteristic diamond peaks in the Raman spectra were at 1 325 $cm^{-1}$ on steel and at 1 322 $cm^{-1}$ on Mo. The spectra are shown in Figure 5.19.

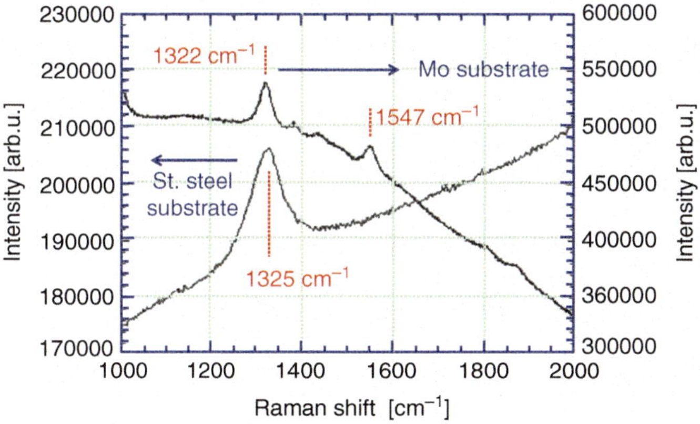


**Figure 5.19** Comparison of the Raman spectrum of the nanocluster diamond coatings on stainless steel (Figure 5.18) with the Raman spectrum of the coating on the Mo substrate.

**Figure 5.20** The SEM image of $MoO_3$ formations on both surfaces of the Mo substrate (a). Diamond aggregates were grown only on the plasma side, preferentially on the $MoO_3$ rods (b).

During diamond PE CVD, a rapid molybdenum oxidation was observed, surprisingly on both sides of the samples. Diamond aggregates were grown on scrambles of $MoO_3$ rods, see Figure 5.20.

Simultaneous oxidation on both surfaces of the Mo plate, including even the bottom side laying on the holder, was surprising. The molybdenum trioxide ($MoO_3$) created on the Mo substrates during PE CVD acquired also bizarre forms, see Figure 5.21. The $MoO_3$ has a melting point of 795°C and it is stable

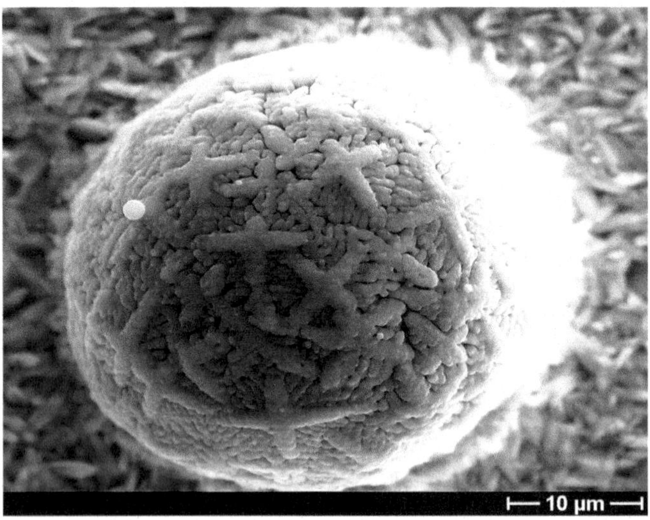


**Figure 5.21** The SEM image of bizarre creation of the molybdenum trioxide grown in PE CVD.

**Figure 5.22** The SEM of the nano-diamonds on the stainless steel substrate (a). Detailed SEM image of the diamond nanocrystals (b).

solid up to 550°C. The observed effect and material can be used in many applications, e.g. as an industrial catalyst, where the H-HEAD treatments can be used for fabrication of large catalytic sheaths or walls.

Figure 5.22 shows 30 nm nanocrystalline diamond grains grown on stainless steel.

The deposition of diamond crystals at atmospheric pressure was already reported also in Chemical Vapor Depositions (CVDs) in simple oxyacetylene flames from conventional welding torch, see Ref. [5.48]. The diamond there was grown on the Mo substrates pretreated by diamond paste and held at a temperature between 700 and 950°C in the $O_2 + C_2H_2$ mixture flowing at about 300 slm. The PE CVD regime presented in H-HEAD, with small flows of air and alcohol vapor and without any need for pretreatments of substrates, is considerably simpler. The possible explanation of the PE CVD regime in the H-HEAD can be based on two factors: (1) The used temperature was under 600°C, where diamond should stay stable without being graphitized, and (2) the graphite is oxidized (vaporized) about 100 times faster than the diamond, which could help in cleaning of the growing diamond from the graphite parts by the reactive oxygen atoms formed in the H-HEAD air plasmas.

## 5.3 Microwave Treatment of Diesel Exhaust

The pulsed microwave plasma was tested in treatments of the particulate matter formed in the diesel exhaust. An idea to use microwave plasma for the removal of particles from cars was patented by G. Lucas and A.I. Al-Shamma'a, see Ref. [5.49]. We have constructed and tested an alternative arrangement with the pulsed microwave power, as shown in Figure 5.23.

To avoid depositions of carbon coatings inside the waveguide from the warm exhaust gas produced by the used diesel engine (2 kW genset), the microwave power line around the plasma was heated to 60°C by an auxiliary water circuit, see Figure 5.23. The diesel exhaust gas was introduced into the tapered waveguide cavity comprised an antenna with a sharp tip positioned in the center of the broader wall of the waveguide, i.e. in the maximum electric field of the dominant mode $TE_{10}$. The sharp tip of the antenna enabled an easy start of the plasma in the system due to the high electric field. After passing the plasma, the treated exhaust gas was analyzed by a special sampling analyzer (impactor) where the content of the carbon particles was pumped out, sorted, and analyzed according particle sizes in comparison with the non-treated exhaust. Figure 5.24 shows a photograph of the test system with the burning plasma, and Figure 5.25 shows the time dependent results of the plasma cleaning.

It was confirmed that the microwave plasma can remove all sizes of the carbon particulate matter from the diesel exhaust gas. However, at the same time, the plasma enriched the treated gas with undesirable nitrogen oxides NO and $NO_2$, as can be seen in Figure 5.26.

The increased repetition frequency represented an increased power and had little effect on the $NO_x$ contents. The main factor was the air plasma, which

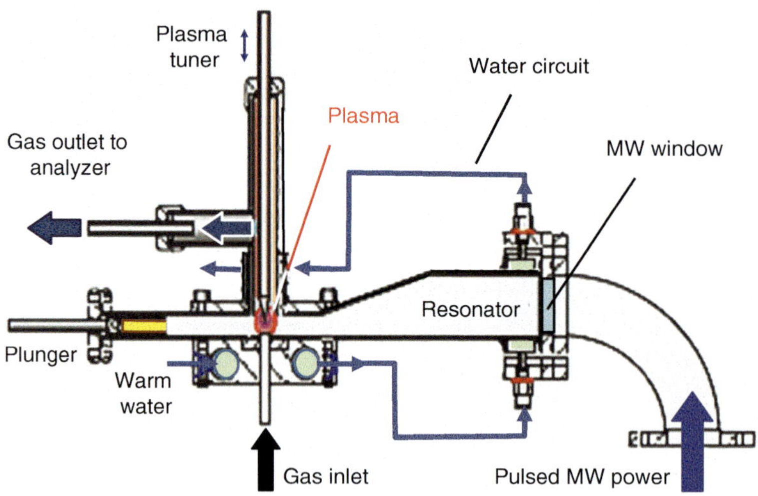


**Figure 5.23** The system with pulsed power of 2.4 GHz, up to 1 kW in 1–20 kHz pulses (duty cycle 1) for testing of removal particulate matter from 2 kW diesel genset exhaust.

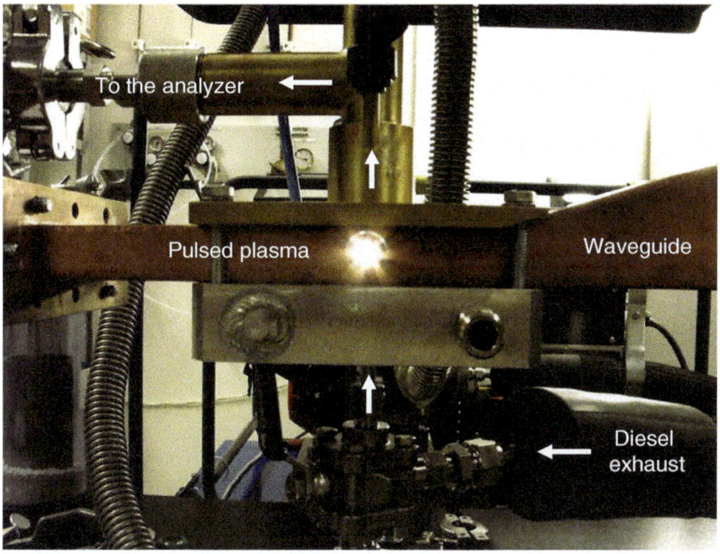


**Figure 5.24** The test system for cleaning of the particulate matter from the diesel exhaust gas. The plasma was observed through a small side window arranged as a quarter-wave cavity.

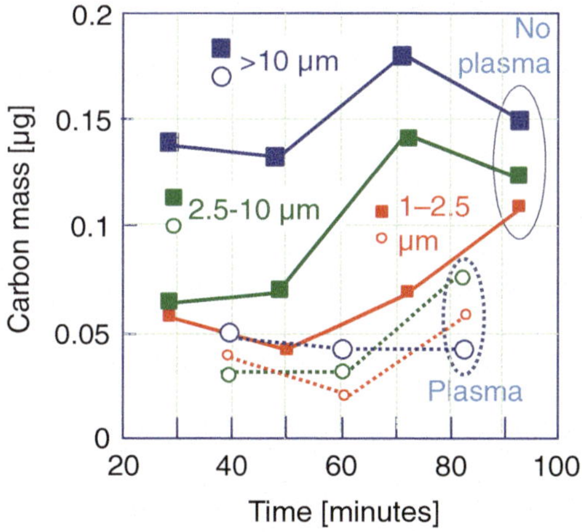


**Figure 5.25** Plasma removal of the particulate matter from the diesel genset. The graph shows differences before and after plasma treatment according to particle size.

generated reactive O atoms. The produced $NO_2$ can be removed by installing a solid catalyst, see Ref. [5.50], and NO can be removed by an aerodynamically stabilized fused hollow cathode plasma, see Ref. [5.51]. However, such cleaning system would be too complex and might reduce the $NO_x$ generation without

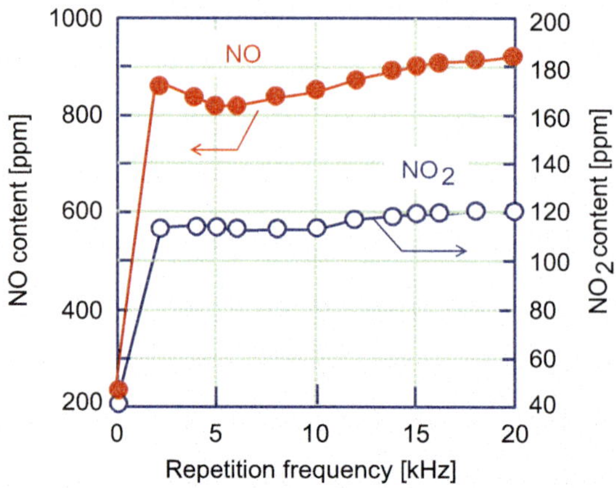


**Figure 5.26** Generation of NO and $NO_2$ in the microwave air plasma-treated diesel exhaust as a function of the repetition frequency of the microwave power pulses.

adding other tools, for example by using nanosecond microwave pulses. This option has not been tested yet.

## 5.4 Microwave Plasma in Liquids

Certain electrode arrangements and reactors for the cold atmospheric plasma can be used for plasma generation in liquids and vapors. Figure 5.27 shows the CAP devices in contact with liquids, above the liquid, under the liquid, or in the vapor produced by liquids. Many such systems are powered by microwaves, see Refs. [5.14, 5.15].

The microwave-generated antenna discharge, in the H-HEAD version and with the antenna without the hollow cathode arrangement, can generate stable plasmas in the water vapor-enriched air, see Ref. [5.56]. Moreover, the antenna can be immersed into the water and generate the plasma partly in the gas bubbles formed by the flowing auxiliary gas. Figure 5.28 shows these regimes. An advantage of the antenna-generated surface wave plasma is the low gas flow rate, as it has been shown above in Figure 5.9. At low flow rates of a plasma gas, the forming of gas bubbles in the liquid is weak, and the plasma generated by the antenna can be stable. Another advantage of the antenna immersed in the liquid is the direct cooling of the nozzle tip by the liquid and the liquid heating, which can be utilized in certain processes.

Another system with the central microwave antenna powering a parallel-plate electrode was tested under water for generation of a broader plasma area, see Figure 5.29. The parallel-plate system generated submerged microwave

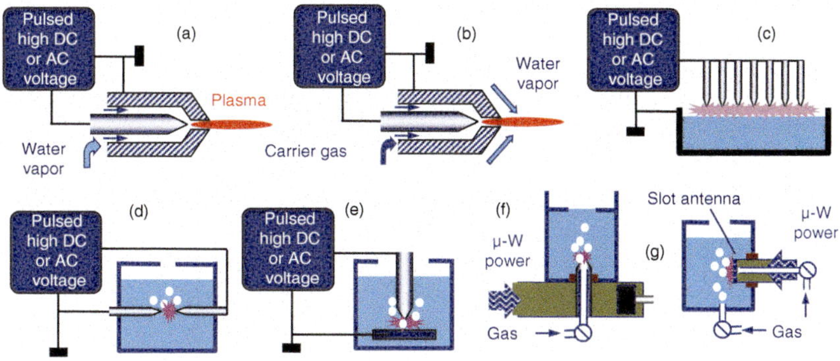


**Figure 5.27** Cold atmospheric plasma (CAP) devices in contact with liquids, or under liquid (water, water mixtures, or other electrolytes). (a) Coaxial electrode system for the plasma jets in water vapor (basic system). (b) System in (a) with the plasma jet in a carrier gas with water vapor admixed at the outlet. (c) A single or multiple electrode system above liquid (electrolyte) in a grounded conductive vessel. (d) Two electrodes submersed in the liquid, based on Ref. [5.52]. (e) An alternative of (d) with a submersed planar grounded electrode. (f) Microwave-powered system with an antenna serving as a gas bubbler, based on Ref. [5.53]. (g) Microwave-powered system with special slot antenna with multiple gas nozzles and an auxiliary gas bubbler for starting or enhancing the plasma bubbles, see Refs. [5.54, 5.55].

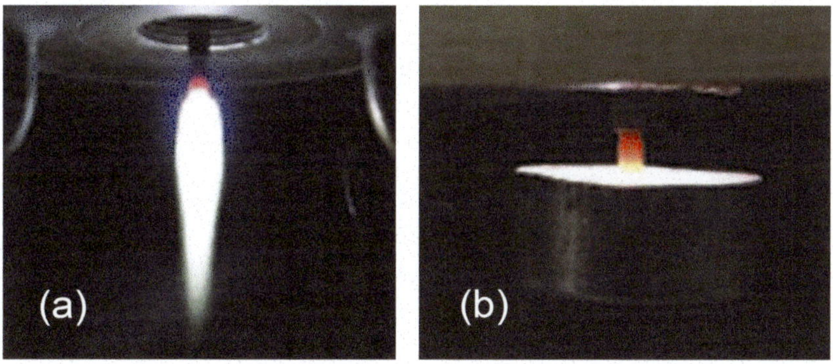


**Figure 5.28** The antenna generated microwave plasma plume in 300 sccm air enriched by the water vapor (a). The microwave antenna with 250 sccm air immersed in the water vessel (b).

plasmas at low microwave power (≈ 20 W). Moreover, the plasma can be generated, with or without bubbling of an auxiliary gas. Because of the low power, the microwave power line was connected to the electrode system inside the reactor vessel by flexible coaxial cable. The connector on the reactor vessel allowed connections of different power sources. In case of the microwave cable, a special adapter was designed and fabricated to couple the power to the ultrahigh voltage (UHV) connector, see Figure 5.30.

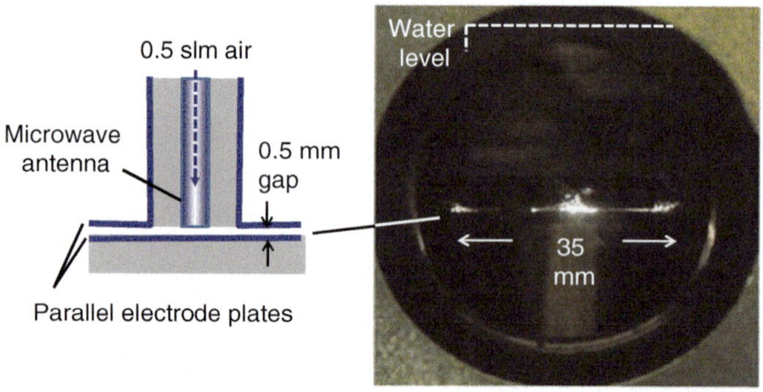


**Figure 5.29** Parallel-plate electrode arrangement with the coaxial microwave antenna for the plasma generation under water. The photograph shows the plasma generated by 2.4 GHz, 80 W incident power, with 16 W reflected power, using bubbles from 0.5 slm (500 sccm) air.

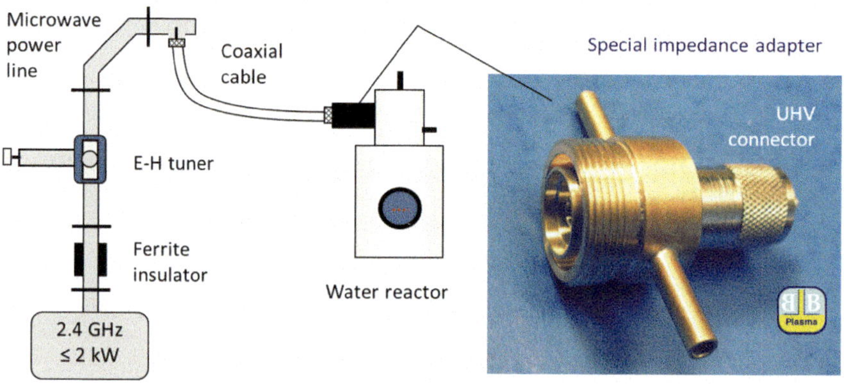


**Figure 5.30** Coupling between the microwave power line and the UHV connector at the reactor.

The discharge treatment of 600 ml tap water was tested in the water reactor to compare effects of different discharges. Three types of discharges were tested: microwave discharge (2.4 GHz, 160 W), RF discharge (13.56 kHz, 200 W), and pulsed DC discharge (2 kHz, 8 ns, 10 W). Comparisons of these treatments for the water temperature after 12 min and for changes of the acidity show the pH factor displayed in Figure 5.31. The results show that all discharges increase the temperature; however, the lowest increase was caused (as can be expected) by the low power short-pulsed DC plasma. The microwave plasma and the RF plasma are similar. However, the pulsed DC plasma caused strong changes in the pH factor compared to the microwave and RF discharges.

Depending on the discharge parameters, on the gas bubbling (if any) and on the electrode arrangements, the discharge treatment changes water properties

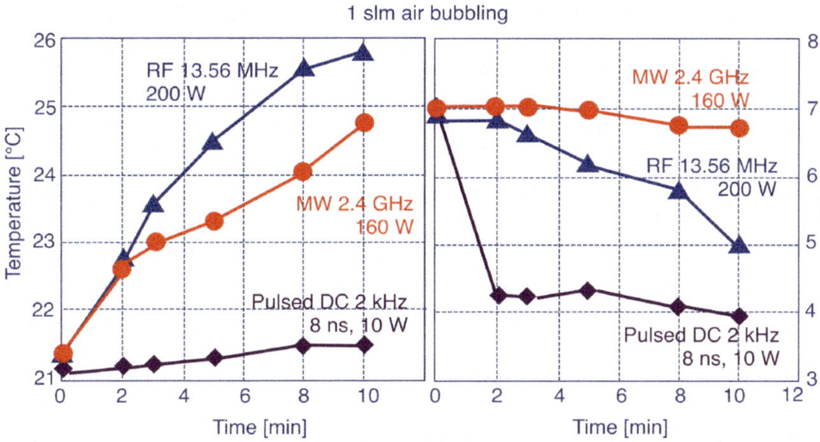


**Figure 5.31** Comparison of three different discharges in treatment of 600 ml of tap water.

and creates plasma-activated water (PAW), plasma-activated tap water (PATW), or plasma-activated liquid medium (PALM). These treatment processes have generated interest because the PAW has confirmed to be efficient in numerous applications. For example, in the activation of agricultural seeds and plants, see Ref. [5.57], for treatment and conservation of food, see Ref. [5.58] for disinfection and antivirus treatments, see Ref. [5.59], or for the therapy of certain kinds of cancer, see Ref. [5.60]. The reader can find numerous references to individual applications, but these applications are beyond the scope of this book. Figure 5.31 shows typical activation process of tap water by three kinds of discharges (heating and changing the acidity given by pH value). According to Refs. [5.57, 5.58], due to acid/base equilibria and the increase of water temperature during plasma activation, chemical species are redistributed from their basic to acid forms followed by decreasing of the pH, as seen in Figure 5.31. Plasma-enhanced dissociations of water molecules and air bubbles form reactive oxygen species (ROS), reactive nitrogen species (RNS), and reactive oxygen and nitrogen species (RONS) forming hydrogen peroxide ($H_2O_2$) and nitric oxide (NO), which are considered important in the activity of the plasma treated / activated water (PAW) in applications.

## 5.5 Microwave Plasma Interactions with Flames

An interesting part of CAP applications is the interaction with the combustion flames. For a long time, there was no general acceptance of the flames as a form of plasma. However, recent measurements by double Langmuir probes in a premixed methane flame revealed even as much as $10^{11}$ $cm^{-3}$ ion density, see Ref. [5.61]. Therefore, an interaction of the non-equilibrium (cold) gas discharges with the flame can generate reactive radicals and plasma-chemical reactions to stabilize combustion processes, to increase the flame extinction limits, and to affect the flame geometry. A consequence could be an improved overall performance of the combustor and stabilization of flames during transients. This can be important in gas turbines, power stations, or aircraft engines for reduction of the emissions.

The reduction of emissions can be achieved by lowering the temperature in the combustor, e.g. in the lean regimes with more air mixed in the fuel. The effect of lower temperature on the pollutions and requirements for optimizations of the flame intensity and efficiency were emphasized in 1971, see Ref. [5.62], but the lowering of the flame temperature usually led to slower chemical reactions and, consequently, to flame instabilities and extinctions. It was confirmed earlier that some parts of the flames can be ionized by certain plasma-chemical reactions to stabilize the flame, see Ref. [5.63]. Therefore, it is important to study how the plasma can affect the flame geometry, the location of an optimal position of the plasma source, see Ref. [5.64], or how to control the discharge effects.

A frequent requirement in the gas turbines is the flat fronts of the flames to increase an interaction efficiency of the flame in the combustor. Experimental tests of the microwave antenna plasma on the shape of flames in the liquid petroleum gas (LPG) revealed the possibility to broaden the flames and increase the flame stability. The LPG represents mixes of propane ($C_3H_8$) and butane ($C_4H_{10}$) usually used as a fuel in heating appliances, cooking equipment, and vehicles.

For tests of the effects of the CAP, several non-conventional plasma arrangements were designed and used by this book's authors, see Ref. [5.65]. Figure 5.32 shows the arrangement with the microwave antenna, with antenna generating a plasma jet in an air flow of 0.5 slm and affecting the flame shape considerably. Without plasma, the LPG flame was stable below about 5 slm, and above this flow, the flame always extinguished. However, after starting the microwave plasma, the shape of the flame changed substantially and increasing the LPG flow rate up to about 8 slm remained possible while keeping the stable flame. Moreover, the front part of the flame was broadened to a flower shape, with branches following four LPG nozzles used in this arrangement.

The temperature in the microwave plasma plume in air was measured by the B-type thermocouple in the axis of the microwave air plasma jet as shown in Figure 5.33. In the measurements, the microwave power delivered to the 0.5 slm air jet was 360 W in the continual wave (CW) regime. The measurements were slightly affected by spatial instabilities of the flames and by strong heating of rhenium cartridge (melting point is 3186°C) of the thermocouple,

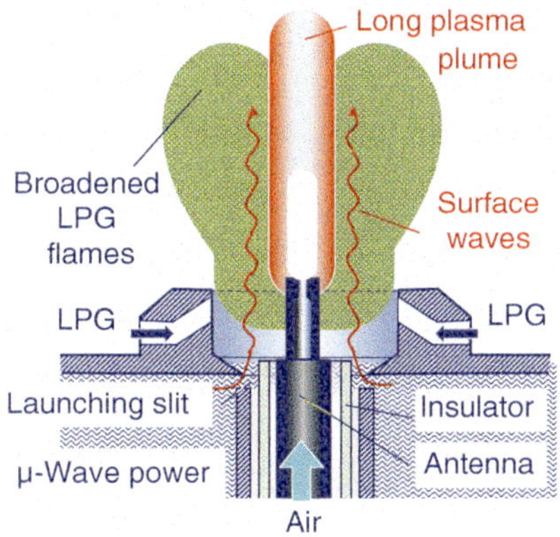


**Figure 5.32** Schematic representation of experimental arrangement for testing the effects of microwave antenna plasma on the LPG flames.

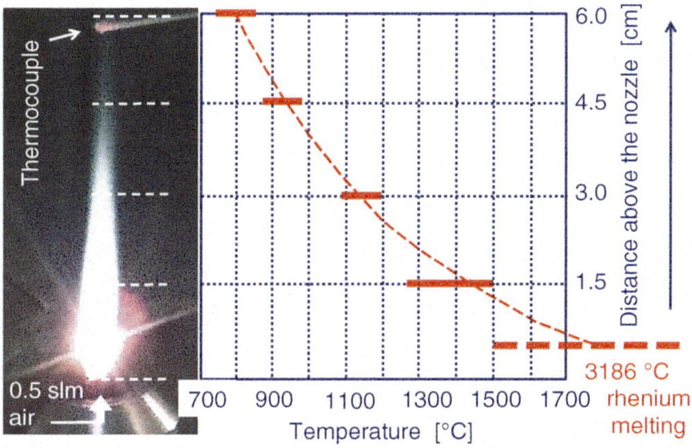


**Figure 5.33** The temperature of the microwave air plasma jet at different distances from the nozzle on the microwave antenna. The temperature near to the nozzle was close to the melting point of rhenium cartridge of the thermocouple (3186°C).

particularly near the nozzle. In Figure 5.33, the heated tip of the thermocouple cartridge is above the plasma plume. The measurements near the microwave antenna were affected by the electric interactions between the microwave antenna and the electrically grounded cartridge of the thermocouple.

Figure 5.34 shows the effects of the air plasma jet on the flame shape. As mentioned above, the LPG flame was stable below flow rates of 5 slm. Above this flow, the flame extinguished, probably because of the mixing of the flames with the central flow of neutral air from the antenna jet, which could exceed the extinction limit for the lean regime. After switching on the microwave power, the shape of the flame changed, and it was stable up to about 8 slm of

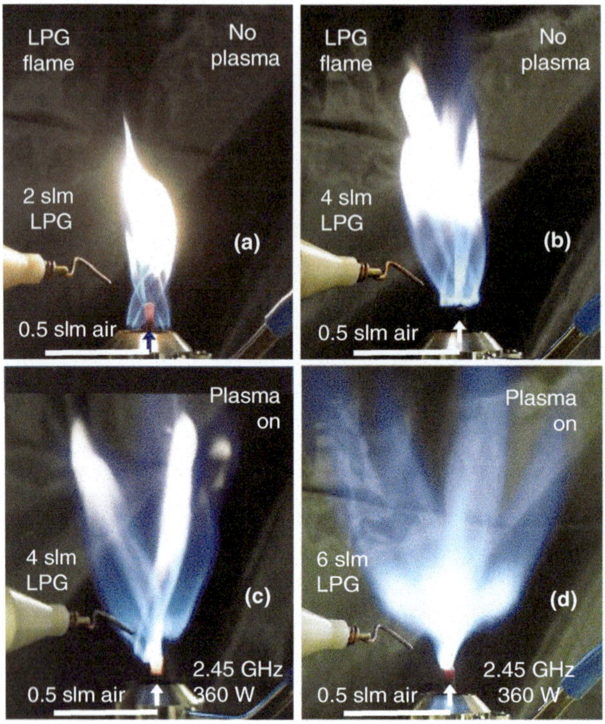


**Figure 5.34** The LPG flames with and without the microwave air plasma jet. (a) The flame at low LPG flow rate. (b) The LPG flame close to the extinction. (c) Stabilizing and shaping effect of the microwave plasma jet on the LPG flame. (d) Stable and broad LPG flame supported by the central microwave air plasma jet.

the LPG forming the flower flames. Figure 5.34 shows the comparison of the individual phases described above.

The temperature measurements of the LPG flame were performed with and without the plasma jet, keeping the 0.5 slm of the central air jet stream in both cases. The thermocouple was positioned at six different distances above the nozzle tip at the microwave antenna, and the temperature was recorded at several radial distances from the axis to find the temperature profiles after broadening the flame by the plasma. The results are shown in Figure 5.35. The temperature of the LPG flame with an axial flow of 0.5 slm of neutral air is shown in Figure 5.35a. The LPG flow rate was kept at 4 slm to avoid the extinction instabilities but was high enough to elevate the flame above the air nozzle. This explains a lower temperature value measured in the axis at 1.5 cm from the nozzle than that at 3 cm and 4.5 cm, where the temperature reached about 1200°C. After switching on the microwave power, the flame stabilized

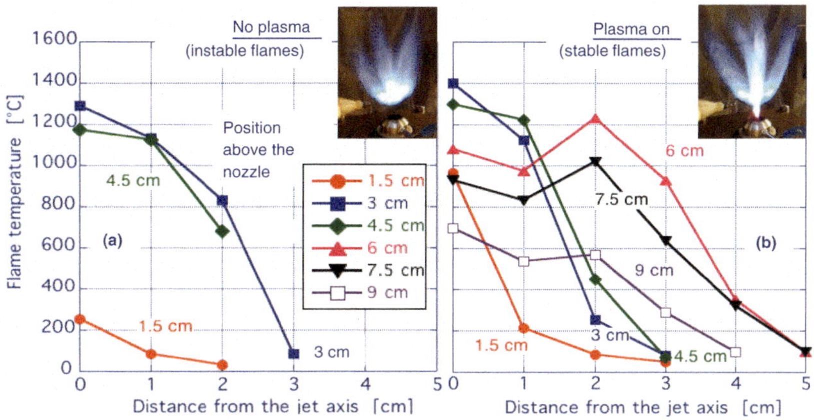


**Figure 5.35** Temperature profiles across the flames at different distances from the antenna tip. (a) The LPG flame in 4 slm with the central neutral air stream of 0.5 slm without plasma. (b) The branched flame in 6 slm LPG after switching-on the power in the 0.5 slm air plasma jet.

and acquired a broader shape with side branches in this bouquet-shaped flame. Consequently, the temperature field was wider, and the temperature of the side branches was comparable with the axial parts of the flame. This can be seen in Figure 5.35b.

The performed experimental tests confirmed possibility of controllable shaping of the flame geometry. Moreover, the plasma assistance confirmed earlier findings of an increased stability of the flames. Even though the performed experiments had a phenomenological character, the obtained results could help in new designs of efficient auxiliary plasma units, e.g. for gas turbines or for aircraft engines.

## 5.6 Microwave Plasmas at Very High Pressures

We haven't mentioned so far the microwave plasma at pressures higher than 1 bar because the simple atmospheric plasma is of the major use in applications. However, several experiments in a hermetic vessel have proven the possibility to operate the plasma at higher than atmospheric pressures. For example, the experiments with the submerged microwave plasma in water (described in Figure 5.29) were tested up to about 1.1 bar. Slightly enhanced pressures can be expected in the microwave plasma jets at high gas flow rates. The pressures as high as 5 bar were already tested successfully for the direct microwave plasma ignitions of gasoline fuel in the combustion engines, see Ref. [5.66]. Even the

30 bar pressure in the mixture of He + $CH_4$ was used in the microwave plasma production of ultrafine carbon particles in a specially designed semi-coaxial 2.45 GHz resonator, see Ref. [5.67].

## References

[5.1] R. Sankaranarayanan, B. Pashaie, and S.K. Dhali: "Characteristics of a barrier discharge in monatomic and molecular gases"; Appl. Phys. Lett. **74** (1999) 3119–3121.

[5.2] L. Bardos and H. Baránková: "Plasma processes at atmospheric and low pressures"; Vacuum **83** (2009) 522–527.

[5.3] H. Baránková and L. Bardos: "Hollow cathode cold atmospheric plasma source with monoatomic and molecular gases"; Surf. Coat. Technol. **163–164** (2003) 649–653.

[5.4] Y.-M. Yang, C.-W. Yuan, and B.-L. Qian: "Measurement of S-band microwave gas breakdown by enhancing the electric field in a waveguide"; IEEE Trans. Plasma Sci. **40** (12) (2012) 3427–3432.

[5.5] B.M. Song, D.A. Hammer, C. Golkowski, and Y.-L. Tian: "Initiation of microwave-induced electrical breakdown of high-pressure gases"; IEEE Trans. Plasma Sci. **31** (1) (2003) 146–156.

[5.6] G.C. Herring and S. Popovic: "Microwave air breakdown enhanced with metallic initiators"; Appl. Phys. Lett. **92** (13) (2008) 131501-1–131501-3.

[5.7] L. Bardos and H. Baránková: "Cold atmospheric plasma: Sources, processes, and applications"; Thin Solid Films **518** (23) (2010) 6705–6713.

[5.8] Z. Károly, G. Kalácska, J. Sukumaran, D. Fauconnier, Á. Kalácska, M. Mohai, and S. Klébert: "Effect of atmospheric cold plasma treatment on the adhesion and tribological properties of polyamide 66 and Poly(Tetrafluoroethylene)"; Materials **12** (**2019**) 658, 1–14. 10.3390/ma12040658.

[5.9] T.S.M. Mui, L.L.G. Silva, V. Prysiazhnyi, and K.G. Kostov: "Surface modification of aluminium alloys by atmospheric pressure plasma treatments for enhancement of their adhesion properties"; Surf. Coat. Technol. **312** (2017) 32–36.

[5.10] G.F. Fernandes, M. Kowalczuk Manosso Amorim, R.G. Turri, et al.: "Plasma treatment of crosslinked polyethylene tubes for improved adhesion of waterbased paints"; Mater. Res. **22** (1) (2019) e20171046, pp. 1–8. http://dx.doi.org/10.1590/1980-5373-MR-2017-1046.

[5.11] A.S. Meshkova, F.M. Elam, S.A. Starostin, et al.: "The role of carrier gas flow in roll-to-roll AP-PECVD synthesized silica moisture barrier films"; Surf. Coat. Technol. **339** (2018) 20–26.

[5.12] K.H. Kale and A.N. Desai: "Atmospheric pressure plasma treatment of textiles using non-polymerising gases"; Indian J. Fibre Text. Res. **36** (11) (2011) 289–299.

[5.13] B. Sharma, R. Kar, A.R. Pala, et al.: "Investigations on the transformation of vertically aligned CNTs to intramolecular junctions by atmospheric pressure PECVD"; Mater. Today Commun. **16** (2018) 178–185.

[5.14] P. Bruggeman and C. Leys: "Non-thermal plasmas in and in contact with liquids"; J. Phys. D: Appl. Phys. **42** (2009) 053001 pp.1–28.

[5.15] Y.A. Lebedev: "Microwave discharges in liquids: fields of applications"; High Temp. **56** (3) (2018) 811–820.

[5.16] S. Jung, H.J. Kim, S. Park, et al.: "The use of atmospheric pressure plasma-treated water as a source of nitrite for emulsion-type sausage"; Meat Sci. **108** (2015) 130–137.

[5.17] N.N. Misra, A. Martynenko, F. Chemat, et al.: "Thermodynamics, transport phenomena, and electrochemistry of external field-assisted nonthermal food technologies"; Crit. Rev. Food Sci. Nutr. (2017) pp. 1–32. http://dx.doi.org/10.1080/10408398.2017.1287660.

[5.18] S.K. Pankaj, H. Shi, and K.M. Keener: "A review of novel physical and chemical decontamination technologies for aflatoxin in food"; Trends Food Sci. Technol. **71** (2018) 73–83.

[5.19] K. Zhang, C.A. Perussello, V. Milosavljevic, et al.: "Diagnostics of plasma reactive species and induced chemistry of plasma treated foods"; Crit. Rev. Food Sci. Nutrit. (2019) pp. 1–23. 10.1080/10408398.2018.1564731.

[5.20] M. Laroussi, X. Lu, and M. Keidar: "Perspective: The physics, diagnostics, and applications of atmospheric pressure low temperature plasma sources used in plasma medicine"; J. Appl. Phys. **122** (2017) 020901, pp. 1–19.

[5.21] S.-H. Choi, W.-S. Jeong, J.-Y. Cha, et al.: "Time-dependent effects of ultraviolet and nonthermal atmospheric pressure plasma on the biological activity of titanium"; Sci. Rep. **6** (11) (2016) 33421. 10.1038/srep.33421.

[5.22] S. Helgadóttir, S. Pandit, V.R.S.S. Mokkapati, et al.: "Vitamin C pretreatment enhances the antibacterial effect of cold atmospheric plasma"; Cell. Infect. Microbiol. **7** (2017) article 43. 10.3389/fcimb.2017.00043.

[5.23] Z. Chen, X. Cheng, L. Lin, and M. Keidar: "Cold atmospheric plasma discharged in water and its potential use in cancer therapy"; J. Phys. D: Appl. Phys. **50** (2017) 015208 (8pp).

[5.24] V. Stepanova, P. Slavicek, J. Kelar,et al.: "Atmospheric pressure plasma treatment of agricultural seeds of cucumber (Cucumis sativus L.) and pepper (Capsicum annuum L.) with effect on reduction of diseases and germination improvement"; Plasma Process. Polym. **15** (2) (2017) 1–9. https://doi.org/10.1002/ppap.201700076.

[5.25] G.J.J.B. De Groot, A. Hundt, A.B. Murphy, et al.: "Cold plasma treatment for cotton seed germination improvement"; Sci. Rep. **8** (2018) 14372. http://dx.doi.org/10.1038/s41598-018-32692-9.

[5.26] R. Molina, C. Lopez-Santos, A. Gomez-Ramirez, et al.: "Influence of irrigation conditions in the germination of plasma treated Nasturtium seeds"; Sci. Rep. **8** (2018) 16442. http://dx.doi.org/10.1038/s41598-018-34801-0.

[5.27] M.M. Kekez, M.R. Barrault, and J.D. Craggs: "Spark channel formation"; J. Phys. D: Appl. Phys. **3** (1970) 1886–1896.

[5.28] S. Kanazawa, M. Kogoma, T. Moriwaki, and S. Okazaki: "Stable glow plasma at atmospheric pressure"; J. Phys. D: Appl. Phys. **21** (1988) 838–840.

[5.29] S. Okazaki, M. Kogoma, M. Uehara, and Y. Kimura: "Appearance of stable glow discharge in air, argon, oxygen and nitrogen at atmospheric pressure using a 50 Hz source"; J. Phys. D Appl. Phys. **26** (1993) 889–892.

[5.30] W. Siemens: "Ueber die elektrostatische Induction und die Verzögerung des Stroms in Flaschendrähten"; Poggendorff's Ann. Phys. Chem. **102** (1857) 66.

[5.31] T. Andrews and P.G. Tait: "VII. On the volumetric relations of ozone, and the action of the electrical discharge on oxygen and other gases"; Phil. Trans. Roy. Soc. (London) **150** (1860) 113–132.

[5.32] G.D. Cristescu and R. Grigirovici: "Optical temperature determinations in the high-frequency "torch" discharge"; Naturwissenschaften **29** (1941) 571–572.

[5.33] L. Mollwo: "Elektronentemperatur und Elektronenrauschen in der hochfrequenten Fackelentladung"; Ann. Physik **7** (1958) 97–129.

[5.34] T. Homola, J. Matousek, M. Kormunda, L.Y.L. Wu, and M. Cernak: "Plasma treatment of glass surfaces using diffuse coplanar surface barrier discharge in ambient air"; Plasma Chem. Plasma Process. **33** (2013) 881–894.

[5.35] L. Bardos and H. Baránková: "Atmospheric Plasma – Yes or No?"; Presented at the 49th TechCon of Society of Vacuum Coaters as the D.M. Mattox Tutorial, April 26, 2006 in Washington DC, published in SVC Summer Bulletin (2006), pp. 42–48.

[5.36] H. Baránková and L. Bardos: "Atmospheric plasma technology for surface treatment"; Galvanotechnik **9** (2004) 2244–2247.

[5.37] L. Bardos and H. Baránková: "Characterization of the cold atmospheric plasma hybrid source"; J. Vac. Sci. Technol. **A23** (4) (2005) 933–937.

[5.38] IPLAS – Innovative Plasma Systems GmbH, Germany, www.cyrannus.com.

[5.39] B.K. Barnes, H. Ouro-Koura, J. Derickson, et al.: "Plasma generation by household microwave oven for surface modification and other emerging applications"; Am. J. Phys. **89** (4) (2021) 372–382. https://doi.org/10.1119/10.0002706.

[5.40] H. Toyota, S. Nomura, and S. Mukasa:: "A practical electrode for microwave plasma processes"; Int. J. Mater. Sci. Appl. **2** (3) (2013) 83–88. 10.11648/j.ijmsa.20130203.12.

[5.41] L. Bardos and H. Baránková: "Device for hybrid plasma processing", Swedish patent No. 9904295-4, priority 1999-11-26, WO0139560 (A1) – 2001- 05-31.US6899054 (B1), JP2003515433 (A).

[5.42] H. Baránková and L. Bardos: "New hybrid source of cold atmospheric plasma"; Surface Coat. Technol. **177–178** (2004) 688–692.

[5.43] H. Baránková, L. Bardos, and D. Soderstrom:: "Cold atmospheric plasma in nitrogen and air generated by the hybrid plasma source"; J. Vac. Sci. Technol. **A24** (4) (2005) 1410–1413.

[5.44] L. Bardos and H. Baránková:: "Cold atmospheric plasma deposition of diamond"; Plasma Proc. Polym. **4** (2007) 533–514.

[5.45] L. Bardos and H. Baránková: "Cold atmospheric plasma treatment of steel buoys for wave energy"; Plasma Proc. Polym. **8** (2011) 658–663.

[5.46] M. Moisan and Z. Zakrzewski:: "Plasma sources based on the propagation of electromagnetic surface waves"; J. Phys. D: Appl. Phys. **24** (1991) 1025–1048.

[5.47] H. Baránková and L. Bardos: "Hollow cathode atmospheric pressure plasma sources for surface treatment"; Surf. Coat. Technol. **174–175** (2003) 63–67.

[5.48] S. Marinkovic and S. Zec:: "Formation of large diamond crystals in oxyacetylene flame chemical vapour deposition"; J. Mater. Sci. **31** (1996) 5999–6003.

[5.49] J. Lucas and A.I. Al-Shamma'a: "Exhaust gas reactor", USA Patent No. 11/233853, September 2005. EP No. 1606497, December 2005. British Patent No. 0302905.5 GB, October 2003.

[5.50] B.M. Penetrante, R.M. Brusasco, B.T. Merritt, and G.E. Vogtlin: "Environmental applications of low-temperature plasmas"; Pure Appl. Chem. **71** (1999) 1829–1835.

[5.51] H. Baránková and L. Bardos: "Effect of the electrode material on the atmospheric plasma conversion of NO in air mixtures"; Vacuum **84** (2010) 1385–1388.

[5.52] O. Takai:: "Solution plasma processing (SPP)"; Pure Appl. Chem. **80** (9) (2008) 2003–2011.

[5.53] S. Nomura, H. Toyota, S. Mukasa, et al.: "Microwave plasma in hydrocarbon liquids"; Appl. Phys. Lett. **88** (2006) 211503 pp.1–3.

[5.54] T. Ishijima, H. Hotta, H. Sugai, and M. Sato: "Multi bubble plasma production and solvent decomposition in water by slot-excited microwave discharge"; Appl. Phys. Lett. **91** (2007) 121501 pp.1–3.

[5.55] T. Ishijima, H. Sugiura, R. Satio, et al.: "Efficient production of microwave bubble plasma in water for plasma processing in liquid"; Plasma Sources Sci. Technol. **19** (2010) 015010 5pp.

[5.56] H. Baránková and L. Bardos: "Atmospheric pressure plasma sources and processing", Chapter 17 in P. Martin, ed.: "Handbook of plasma processing technology", William Andrew Inc., 2009.

[5.57] F. Judée, S. Simon, C. Bailly, and T. Dufour: "Plasma-activation of tap water using DBD for agronomy applications: identification and quantification of long lifetime chemical species and production/consumption mechanisms"; Water Res. **133** (2018) 47–59.

[5.58] B. Adhikari, M. Adhikari, B. Ghimire, et al.: "Cold atmospheric plasma-activated water irrigation induces defense hormone and gene expression in tomato seedlings"; Sci. Rep. **9** (2019) 16080. https://doi.org/10.1038/s41598-019-52646-z.

[5.59] L. Guo, Z. Yao, L. Yang, et al.: "Plasma-activated water: an alternative disinfectant for S protein inactivation to prevent SARS-CoV-2 infection"; Chem. Eng. J. (2020) in print. https://doi.org/10.1016/j.cej.2020.127742.

[5.60] A. Azzariti, R.M. Iacobazzi, R. Di Fonte, et al.: "Plasma-activated medium triggers cell death and the presentation of immune activating danger signals in melanoma and pancreatic cancer cells"; Sci. Rep. **9** (2019) 4099. |https://doi.org/10.1038/s41598-019-40637-z.

[5.61] J. Wild, P. Kudrna, M. Tichy, et al.: "Electron temperature measurement in a premixed flat flame using the double probe method"; Contrib. Plasma Phys. **52** (8) (2012) 692–698. |https://doi.org/10.1002/ctpp.201200005.

[5.62] F.J. Weinberg: "Combustion temperatures: the future?"; Nature **233** (9) (1971) 239–241.

[5.63] A. Von Engel and J.R. Cozens: "Origin of excessive ionization in flames"; Nature **202** (1964) 480.

[5.64] W. Kim, H. Do, M.G. Mungal, and M.A. Cappelli: "Optimal discharge placement in plasma-assisted combustion of a methane jet in cross flow"; Combust. Flame **153** (2008) 603–615.

[5.65] L. Bardos and H. Baránková: "Shaping of the flame geometry by non-conventional cold plasma arrangements"; Plasma Res. Express **2** (2020) 035014 (1–6). https://doi.org/10.1088/2516-1067/abb8e6.

[5.66] K. Linkenheil, H.-O. Ruoß, T. Grau, et al.: "A novel spark-plug for improved ignition in engines with Gasoline Direct Injection (GDI)"; IEEE Trans. Plasma Sci. **33** (5) (2005) 1696–1702. 10.1109/TPS.2005.856409.

[5.67] H. Yagi, T. Ide, H. Toyota, and Y. Mori: "Generation of microwave plasma under high pressure and fabrication of ultrafine carbon particles"; J. Mater. Res. **13** (6) (1998) 1724–1727.

# 6

# New Applications and Trends in the Microwave Plasmas

Based on our experience we can conclude that **the plasma is well-controllable environment-friendly medium utilizing electric power and enabling the very-high temperature processes and low-temperature to very-low-temperature reactions unavailable in other methods**. Therefore, plasma-based technologies are developing so rapidly that new trends are quickly turned to obsolete ones. Nevertheless, several trends and applications, which have broad applicability, will probably stay in use in the future. Some of them were already mentioned in Chapters 4 and 5.

As introduced in Chapter 1, the applications of the microwave power intensified after the World War Two (WWII). Besides new communication technologies, the microwave cavities were soon developed and used for the acceleration of electrons and forming of high-energy electron beams for different applications in microtrons, see Ref. [6.1]. Such acceleration method was proposed in 1944, see Ref. [6.2]. Under high-vacuum conditions, the surfaces of the stainless-steel cavities in accelerators must be perfectly clean, see Ref. [6.3]. To maximize the beam energy, a recent trend is an in-situ high-vacuum cleaning and outgassing in the accelerators using the radio frequency (RF) and microwave plasmas, see Refs. [6.4, 6.5]. The frequency range is typically between 1–3 GHz, and the electronic cyclotron resonance (ECR) arrangements are used. The principles and gases for such cleaning are similar to the processes known in the plasma processing (see Figure 3.3, Chapter 3). These applications are still under development.

The low-pressure microwave plasma is frequently used in Plasma-Enhanced Chemical Vapor Deposition (PE CVD), e.g. in deposition of diamond coatings. High-quality single-crystal diamond coatings can be applied in new types of lasers. In Raman lasers, they enable higher gain, greater power densities, and changeable wavelengths, see Refs. [6.6, 6.7]. The deposition systems for single-crystal diamond coatings utilize resonators similar to that shown in Figure 1.26a, Chapter 1. However, the single-crystal diamond coatings require selected processing parameters and procedures.

Another interesting new application of the microwave plasma at reduced pressure is plasma annealing, see Ref. [6.8]. It offers better effects than any

*Microwave Plasma Sources and Methods in Processing Technology*, First Edition.
Ladislav Bárdoš and Hana Baránková.

conventional heating and has been applied, e.g. in the synthesis of transition metal oxides like tantalum oxide ($Ta_2O_5$) or zinc oxide (ZnO), for upgrading their protective and optoelectronic properties. Several other interesting trends and arrangements exist in the PE CVD of thin films by low-pressure microwave plasma, see Ref. [6.9]. Kind of microwave plasma annealing was presented as a sintering of powder in Figure 5.16, Chapter 5.

Due to frequent requirements of the large-area microwave plasma systems, new concepts were designed and developed for industrial microwave ECR systems, Ref. [6.10], in particular, the distributed ECR arrangements, see Ref. [6.11]. Such systems can be used in different large-area or large-volume plasma applications, e.g. in plasma etching, PE CVD, or in plasma nitriding. The ECR plasma was described in its application for oxidation of silicon (Section 4.1.2, Chapter 4). For the simple illustration the planar plasma arrangement based on the distributed ECR is shown in Figure 6.1.

Novel high-density low-pressure plasma systems based on the microwave surface waves are innovative devices, e.g. in the dielectric etchers for 12 inch and larger wafers in microelectronics, see Ref. [6.12]. These systems do not need the magnetic field. Principles of the wave propagation on the interface plasma-dielectric and several arrangements were published in Ref. [6.13]. The microwave power is delivered directly from the waveguide by small slots onto the dielectric window (usually quartz) on the plasma reactor, or these slots are arranged in a resonator cavity installed on a dielectric window. Another arrangement was designed as a resonator with radiating slits positioned in a cylindrical geometry. The device, called a slot antenna (SLAN), was described in Ref. [6.14]. The authors claimed a better power transfer than in the coils of the Lisitano slow-wave structure, described in Figure 4.1c, d, Chapter 4. Figure 6.2 shows schematic descriptions of the systems. The power delivery from slots in a resonator arrangement was recently used for the atmospheric plasma (see Figure 5.4d, Chapter 5.1).

Although these low-pressure systems can be arranged for large-volume plasmas, their general weak point is the dielectric window. (Some non-conventional arrangements without windows, e.g. with metal antennas shown in Figure 4.47, were described in Section 4.3.2, Chapter 4.) The window has always limited life-time due to the inner structural changes, particularly at high microwave powers. However, more serious practical disadvantages are the limited applications of such systems in depositions of coatings, which absorb or reflect the microwave power, e.g. carbon-based coatings or metal coatings. Therefore, these systems are more useful in dry etching or annealing processes, but also in the PE CVD of Si-O films. These insufficiencies are probably among the main reasons for the development of new technologies using the microwave plasmas at atmospheric pressure, where neither dielectric

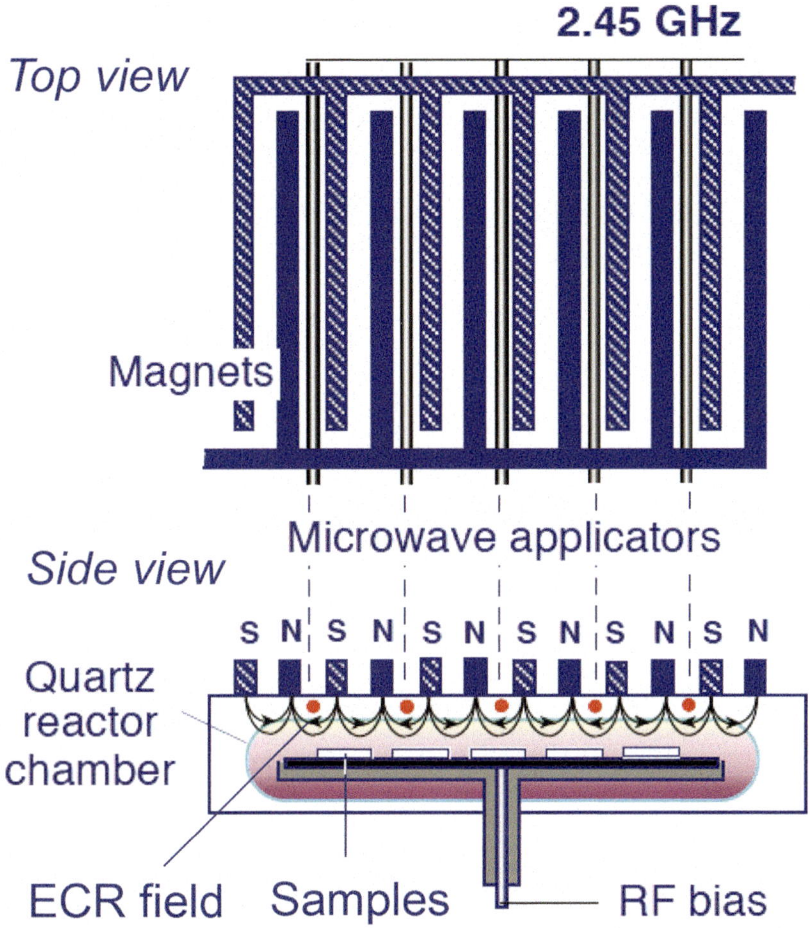


**Figure 6.1** Schematic of planar microwave plasma system utilizing distributed electron cyclotron resonance arrangement. The microwave applicators can be arranged as antennas or primary plasma tubes connecting opposite waveguides. The ECR magnetic induction for 2.45 GHz is $8.76 \times 10^{-2}$ Tesla.

windows nor vacuum pumps are needed. Still, the atmospheric plasmas have serious drawbacks, e.g. the limited outreach of the plasma due to collisions.

Several approaches with the atmospheric plasma, developed and tested in our lab, have been described in Chapter 5. In this chapter, we will point out other interesting abilities in applications of the microwave atmospheric plasma. It was confirmed that the hot microwave arcs can be generated with almost 100% efficiency, with all delivered power absorbed in the plasma plume. Such arcs are generated by absorbing the microwave power in polar molecules.

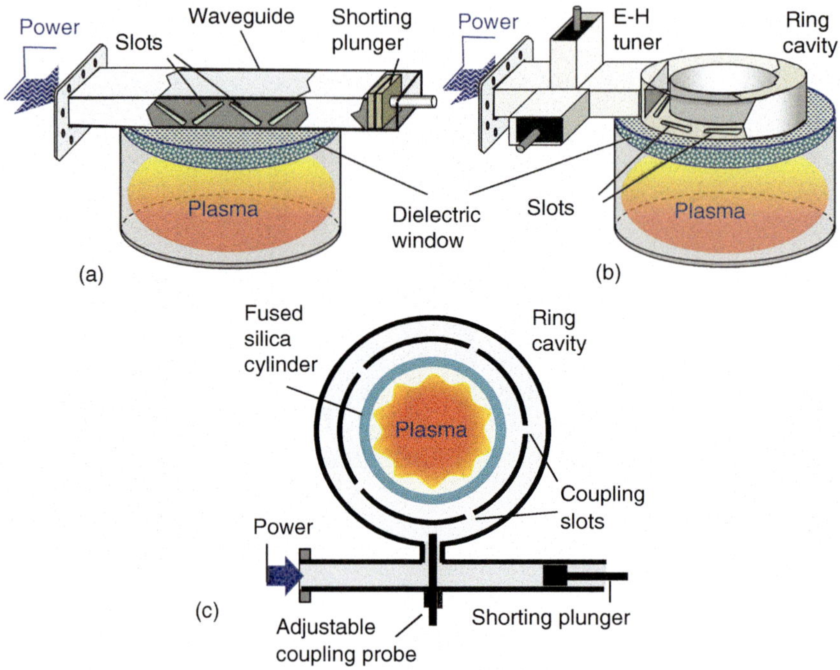


**Figure 6.2** Examples of high-density low-pressure plasma systems based on the microwave surface waves. The top-wall excitation of surface waves from slots in rectangular waveguide (a) and from slots in a ring cavity (b). The side-wall excitation by slot antennas in SLAN system is shown in cross-section (c).

The absorption is based on losses due to an intermolecular friction and high collisions among the molecules, see Ref. [6.15]. The main feature of such microwave plasma arcs is their long uniform high-temperature columns (see, e.g. Figure 5.6, Chapter 5), which is impossible in conventional direct current (DC) or RF arc discharges. The microwave arcs can be used, e.g. in the coal gasification and in gasifying of different wastes at a lower power consumption than in conventional DC and RF electric arcs, see Refs. [6.15, 6.16].

Interesting results were reported on the testing of radioactive decontamination of cobalt oxides by the microwave plasma in $CF_4 + O_2$ gas mixtures (ratio 1:15) at atmospheric pressure, see Ref. [6.17]. The 2.45 GHz 1.5 kW microwave torch was generated by coaxially arranged antenna coupled with the waveguide power line. In 200 seconds, the entire deposit of the radioactive cobalt oxide was removed from the plane surface. Thermodynamic equilibrium calculations on $CF_4 + O_2$ systems containing cobalt explained the mechanism of metal oxide removal by the microwave plasma. Similar calculations on the systems containing actinides (U, Pu, etc.) and lanthanides (La, Gd, etc.) indicated that the removal process can apply to these radioactive elements. More results are reported in the recent handbook, see Ref. [6.18]. These results can lead to the future plasma-based nuclear waste management of medium-level or low-level radioactive waste and replace conventional methods requiring a lot of deactivation procedures and deep storages of the fuel.

An interesting experiment was recently reported, see Ref. [6.19]. The authors proposed a propulsion thruster that utilizes microwave plasma in air. The system is similar to the arrangement shown in Figure 5.4b, Chapter 5, where the discharge tube which passes the tapered waveguide section is closed at the bottom with an RF igniter and an inlet of the pressurized air. Figure 6.3 shows such an arrangement.

The authors used tapered part of the waveguide and a quartz tube with an upper opened outlet. The authors measured the propulsion force with a hollow metal ball placed at the outlet of the reactor tube with pulsed plasma powered by 2.45 GHz and 1 kW. The vibrations of the ball were compensated by a small hole covered by the small metal bead of defined weight. This was used for estimates of the thrust force. The total propulsion pressure was estimated at 2.4 $10^4$ N/m$^2$, which is comparable to that generated by the airplane jet engines using fossil fuels. Such carbon emission-free thruster could potentially be used as a jet thruster. The principle of this experiment can be explained by the general law for an ideal gas, see Eq. (3.10) in Section 3.3.1, Chapter 3. Considering that the temperature of the gas in each pulse igniting the plasma instantly peaks to about 10 000 °C, the corresponding pressure increase from the normal value (1 bar) at about 50°C will grow 200 times. It could be important to study possible by-products generated by this system, like NO and $NO_2$, similar to

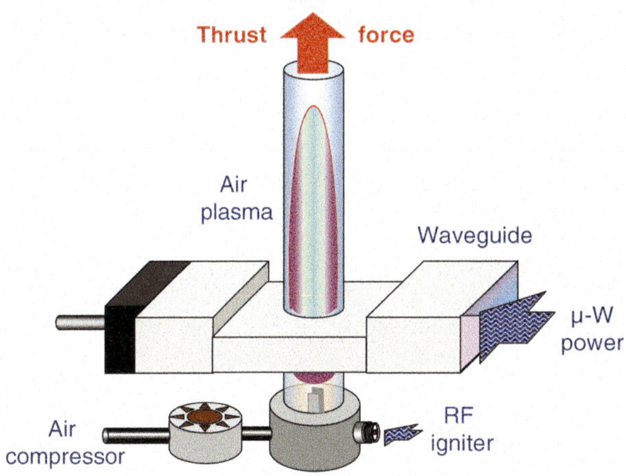


**Figure 6.3** An illustration of the experimental system for testing the thrust force of the atmospheric microwave air plasma.

those found in the microwave cleaning of diesel exhaust (see Figure 5.26, Chapter 5), to avoid environmental pollution.

The production of NO in microwave systems similar to Figure 6.3 was recently used for the NO enrichment of water, which was successfully applied as a supportive medium in treatment of several types of cancer, see Ref. [6.20]. The method was examined both, in vitro and in vivo. It is important to emphasize that there are many reports about successful medical treatments by the cold atmospheric plasma, including cancer. There are special conferences and conference sessions about this field, called *plasma medicine. The plasma-activated water* (PAW) and their applications (including cancer treatment) was mentioned in Chapter 5, section 5.4. This is an exciting field of research and applications for the future.

Many applications of the atmospheric microwave plasma are in the field of energy and environment, e.g. in the plasma-assisted production of hydrogen, applicable in fuel cells or in fuel enrichments in engines. The microwave plasma was effective in the *hydrogen production* by the plasma conversion of methane, see Ref. [6.21]. Another way to produce the hydrogen-rich gas, like synthesis gas $H_2$ + CO, is to use the microwave plasma in the mixture of water with ethanol, see Ref. [6.22]. This mixture was used in other types of discharges (see, e.g. Ref. [6.23]) because the bioethanol is accessible as a waste product. The tested microwave system is simple, as illustrated in Figure 6.4. For more details, see, e.g. Refs. [6.22, 6.24].

As introduced in Section 5.4, Chapter 5, a great future can be expected in different methods utilizing the microwave plasma above, at, and in the liquids. Many references, details of different systems, and applications are available, see Refs. [6.25, 6.26, 6.27]. The main applications are, e.g. in new environmental methods for cleaning of the wastewater and in producing hydrogen-based

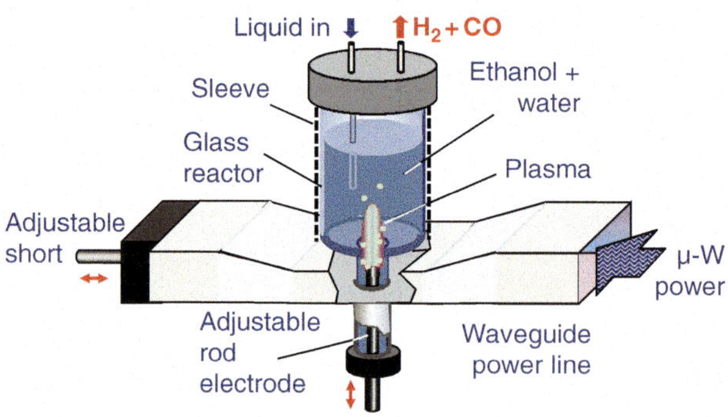


**Figure 6.4** Illustration of the experimental system for production of hydrogen-rich gas.

fuels, Ref. [6.28], producing nanoparticles (NPs), and in several other plasma-chemical reactions applicable in healthcare, the food industry, power production, etc.

The production of NPs is an important part in many applications in the material science, medicine, etc. The microwave plasma has been found useful in this field. According to the review, see Ref. [6.29], the nanomaterials have been intensively developed because of their unique electrical, optical, magnetic, and catalytic properties. The microwave plasma in liquid electrolytes can be used to produce metal NPs. The process consists of three phases:

1) Generation of atoms or small clusters from precursors, such as metal salts, metal oxides, metal complexes
2) Formation of nuclei from atoms or clusters
3) Controlled growth of nuclei to NPs

For example, in Ref. [6.30], Au/Pd alloy NPs were synthesized by the microwave plasma in liquid process using $HAuCl_4$, $H_2PdCl_4$, and $\alpha$-thioglycerol as precursors and a stabilizer. The experimental arrangements are arranged with the waveguide-powered coaxial microwave antennas exciting plasma in a liquid electrolyte through a dielectric window. Alternative microwave arrangements with the microwave plasma in liquids for development of NPs are described in Ref. [6.31]. In this reference, the authors propose microwave systems to treat wastewater.

Another interesting application of the microwave atmospheric plasma is the synthesis of graphene, see Ref. [6.32]. Graphene is an attractive material with the two-dimmensional honeycomb arrangement of carbon atoms. It is considered the world's strongest light material which can be added to other materials to enhance their strengths. It has other unique properties and applications (https://www.graphene-info.com/graphene-applications). The graphene was synthesized using an atmospheric microwave plasma jet, consisting of a mixture of ethanol vapor in argon. Figure 6.5 shows the arrangement.

Because the graphene synthesis from ethanol vapor required controlled temperatures, the end part of the plasma column was cooled or heated by an external heat exchanger. The graphene material was synthesized on membranes of a filter sampler. The microwave plasma jet was generated by the waveguide Surfatron (Surfaguide) arrangement described in Sections 4.2, 4.3, 4.3.1 (Chapter 4) and Figure 5.2 (Chapter 5).

As explained in Section 2.3, Chapter 2, the plasma has characteristic critical density $n_{crit}$ dependent on the frequency f of the used generating power $n_{crit} = 1.24 \times 10^{-2} f^2$, see Eq. (2.10). The plasma with density higher than critical density will reflect the power; therefore, the higher the frequency, the higher the density generated. The authors of the article [6.24] compared two microwave

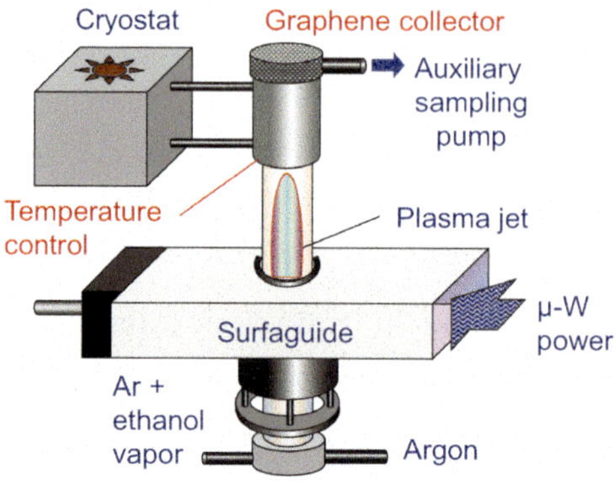


**Figure 6.5** An illustration of the experimental system for the graphene synthesis from ethanol vapor in the atmospheric microwave plasma generated by the waveguide Surfatron.

frequencies 2.45 GHz and 5.8 GHz. They tested several gases and they concluded the electron density in the atmospheric plasma jet for the same absorbed microwave power is almost two times higher at the higher frequency. Therefore, an expected future trend can be a utilization of higher frequencies. As described in Section 1.2, Chapter 1, several types of high-power and high-frequency microwave generators are available, see Ref. [6.33], e.g. the Gyrotron generator, which is a vacuum tube in a strong magnetic field utilizing ECR in a high-power linear-beam to generate millimeter-wave electromagnetic waves in the frequency range from about 20 to 527 GHz. For details, see, e.g. Ref. [6.34]. Of course, the ECR magnetic field for the frequency of 500 GHz is high ($B = f\, m_e/e \approx 2.8$ T = 28 000 Gauss) and requires superconducting magnets. Therefore, such generators are applicable in big Tokamak fusion devices. Lower-frequency klystrons or travelling wave tube (TWT) generators are available for laboratory experiments and future developments but only in well-shielded electromagnetic arrangements.

## References

[6.1] R.E. Jennings: "The microtron"; Contemp. Phys. **2** (4) (1961) 277–285. https://doi.org/10.1080/00107516108202658.

[6.2] V.I. Veksler: "New method of accelerating relativistic particles"; (in Russian) Doklady Akad. Nauk SSSR **43** (8) (1944) 346–348.

[6.3] K.J. Middleman, J.D. Herbert, and R.J. Reid: "Cleaning stainless steel for use in accelerators – Phase 1"; Vacuum **81** (2007) 793–798. https://doi.org/10.1016/j.vacuum.2005.11.057.

[6.4] M. Doleans, P.V. Tyagi, R. Afanador, et al.: "In-situ plasma processing to increase the accelerating gradients of superconducting radio-frequency cavities"; Nucl. Instr. Methods Phys. Res. A **812** (2016) 50–59.

[6.5] P. Berrutti, B. Giaccone, M. Martinello,et al.: "Plasma ignition and detection for in-situ cleaning of 1.3 GHz 9-cell cavities"; J. Appl. Phys. **126** (2019) 023302. https://doi.org/10.1063/1.5092235.

[6.6] S. Antipov, A. Sabella, R.J. Williams, et al.: "1.2 kW quasi-steady-state diamond Raman laser pumped by an $M^2 = 15$ beam"; Opt. Lett. **44** (10) (2019) 2506–2509.

[6.7] G.J. Varnau and H. deWit: "Diamond Raman lasers offer multifaceted potential"; Photonics Spectra **9** (2020) 1–7. https://www.photonicsspectra-digital.com/photonicsspectra/septembe...ePagedArticle.action?articleId=1616337&app=false#articleId1616337.

[6.8] R. Ramadan, J.G. Simiz, M.D. Ynsa, and M.M. Silvan: "Microwave plasma annealing of sol-gel deposited tantalum oxide and zinc oxide films"; Vacuum **149** (2018) 336–342. https://doi.org/10.1016/j.vacuum.2018.01.007.

[6.9] J. Musil: "Deposition of thin films using microwave plasmas: present status and trends"; Vacuum **47** (2) (1996) 145–155.

[6.10] C.M. Ferreira and M. Moisan, editors: "Microwave discharges: fundamentals and applications", Plenum Press, New York, 1993.

[6.11] J. Pelletier, A. Lacoste, Y. Arnal, et al.: "New trends in DECR plasma technology: Applications to novel duplex treatments and process combinations with extreme plasma specifications"; Surface Coat. Technol. **139** (2001) 222–232.

[6.12] H. Kokura, S. Yoneda, K. Nakamura, et al.: "Diagnostic of surface wave plasma for oxide etching in comparison with inductive RF plasma"; Jpn. J. Appl. Phys. **38** (9A) (1999) 5256–5261.

[6.13] H. Sugai, I. Ghanashev, and M. Nagatsu: "High-density flat plasma production based on surface waves"; Plasma Sources Sci. Technol. **7** (1998) 192–205.

[6.14] F. Werner, D. Korzec, and J. Engemann: "Slot antenna 2.45 GHz microwave plasma source"; Plasma Souces Sci. Technol. **3** (4) (1994) 473–481.

[6.15] V. Jain, A. Visani, C. Patil, et al.: "Gliding arc triggered microwave plasma arc at atmospheric pressure for coal gasification application"; Int. J. Mod. Phys. **32** (2014) 1460345 1–8. https://doi.org/10.1142/S2010194514603457.

[6.16] EE Publishers (https://www.ee.co.za): "Plasma gasification of waste: Arcs and microwaves offer advantages"; Sept. (2020) 49 pp. 1–7.

https://www.ee.co.za/article/plasma-gasification-waste-arcs-microwaves-offer-advantages.html.
[6.17] H.F. Windarto, T. Matsumoto, H. Akatsuka, and M. Suzuki: "Decontamination process using $CF_4$-$O_2$ microwave discharge plasma at atmospheric pressure"; J. Nucl. Sci. Technol. **37** (2000) 787–792.
[6.18] R. Kar, A. Bute, N. Chand, et al.: "Removal of radioactive waste by nonthermal plasma etching: trends for the promising future"; Chapter 20 in "Handbook of environmental materials management", C.M. Hussain (ed.), Springer Nature, Switzerland, AG, 2020. https://doi.org/10.1007/978-3-319-58538-3_224-1.
[6.19] D. Ye, J. Li, and J. Tang: "Jet propulsion by microwave air plasma in the atmosphere"; AIP Adv. **10** (2020) 055002 (1–4). https://doi.org/10.1063/5.0005814.
[6.20] C.B. Lee, I.H. Seo, M.-W. Chae, et al.: "Anticancer activity of liquid treated with microwave plasma-generated gas through macrophage activation"; Hindawi Oxid. Med. Cell. Longevity (**2020**) 2020 Article ID 2946820, 13 pages. https://doi.org/10.1155/2020/2946820.
[6.21] M. Jasinski, M. Dors, and J. Mizeraczyk: "Application of atmospheric pressure microwave plasma source for production of hydrogen via methane reforming"; Eur. Phys. J. D **54** (2009) 179–183. https://doi.org/10.1140/epjd/e2008-00221-1.
[6.22] X. Zhao, B. Sun, T. Zhu, et al.: "Pathways of hydrogen-rich gas produced by microwave discharge in ethanol-water mixtures"; Renewable Energy **156** (2020) 768–776. https://doi.org/10.1016/j.renene.2020.04.088.
[6.23] L. Bardos, H. Barankova, and A. Bardos: "Production of hydrogen-rich synthesis gas by pulsed atmospheric plasma submerged in mixture of water with ethanol"; Plasma Chem. Plasma Process. **37** (2017) 115–123. https://doi.org/10.1007/s11090-016-9766-6.
[6.24] B. Sun, X. Zhao, Y. Xin, and X. Zhu: "Large capacity hydrogen production by microwave discharge plasma in liquid fuels ethanol"; Int. J. Hydrogen Energy **42** (2017) 24047–24054. http://dx.doi.org/10.1016/j.ijhydene.2017.08.052.
[6.25] H.-E. Porteanu, D. Wolf, and W. Heinrich: "Frequency dependence of the capacitive excitation of plasma: an experimental proof"; J. Appl. Phys. **127** (2020) 183301 (1–9). https://doi.org/10.1063/5.0005045.
[6.26] Y.A. Lebedev: "Microwave discharges in liquid dielectrics"; Plasma Phys. Rep. **43** (6) (2017) 685–695.
[6.27] Y.A. Lebedev: "Microwave discharges in liquids: fields of applications"; High Temp. **56** (5) (2018) 811–820.
[6.28] A. Hamdan, J.-.L. Liu, and M.-S. Cha: "Microwave plasma jet in water: characterization and feasibility to wastewater treatment"; Plasma

Chem. Plas. Process. **38** (2018) 1003–1020. https://doi.org/10.1007/s11090-018-9918-y.

[6.29] T. Yonezawa, D. Čempel, and M.T. Nguyen: "Microwave-induced plasma-in-liquid process for nanoparticle production"; Bull. Chem. Soc. Jpn. **91** (2018) 1781–1798. https://doi.org/10.1246/bcsj.20180285.

[6.30] H. Shirai, M.T. Nguyen, D. Čempel, et al.: "Preparation of Au/Pd bimetallic nanoparticles by a microwave-induced plasma in liquid process"; Bull. Chem. Soc. Jpn. **90** (2017) 279–285. https://doi.org/10.1246/bcsj.20160333.

[6.31] S. Horikoshi and N. Serpone: "In-liquid plasma: A novel tool in the fabrication of nanomaterials and in the treatment of wastewaters"; Royal Soc. Chem. Adv. **7** (2017) 47196 47218. https://doi.org/10.1039/c7ra09600c.

[6.32] E. Tatarova, J. Henriques, C.C. Luhrs, et al.: "Microwave plasma based single step method for free standing graphene synthesis at atmospheric conditions"; Appl. Phys. Lett. **103** (2013) 134101. https://doi.org/10.1063/1.4822178.

[6.33] R.S. Symons: "Modern microwave power sources"; IEEE AESS Sys. Mag., January (2002) 19–26.

[6.34] N.G. Sabri: "Production of high power by using gyrotron device for electron cyclotron resonance heating in tokamak reactor"; Energy Procedia **18** (2012) 944–953.

# 7

# Appendices

## 7.1 List of Symbols and Abbreviations

### Chapter 1

λ – wavelength in metric units.

c – c = 3 x $10^8$ m/s – the velocity of light (universal constant). It is c = $(\mu_0 \varepsilon_0)^{-1/2}$, where the vacuum permeability $\mu_0$ = 1.256 x $10^{-6}$ H/m = 4π $10^{-7}$ H/m (Henry H = kg $m^2$ $s^{-2}$ $A^{-2}$) and the vacuum permittivity $\varepsilon_0$ = 8.85 x $10^{-12}$ F/m (Farad F = $A^2s^4$ $kg^{-1}$ $m^{-3}$).

f – frequency of the wave (f = c/λ). Units are Hertz (Hz), 1 Hz = 1/s. kHz = $10^3$ Hz (Kilohertz), MHz = $10^6$ Hz (Megahertz), GHz = $10^9$ Hz (Gigahertz), THz = $10^{12}$ Hz (Terahertz), PHz = $10^{15}$ Hz (Petahertz), EHz = $10^{18}$ Hz (Exahertz), ZHz = $10^{21}$ Hz (Zettahertz), YHz = $10^{24}$ Hz (Yottahertz).

**E** or ***E*** – Vector of the electric field. Units are volt/meter E = V/m.

**B** or ***B*** – Vector of the magnetic induction. Units are Tesla = kg/($As^2$) = V.s/$m^2$ and 1 T = $10^4$ Gauss.

**H** or ***H*** – Intensity vector of the magnetic field. Units are A/m. H = B/μ (μ is the magnetic permeability of the medium).

$TE_{mn}$, $TM_{mn}$ – Transversal electric or magnetic modes in the microwave waveguides. Indices m and n show numbers of half-wavelengths in Cartesian coordinates x and y in the waveguide.

$TE_{10}$ – Dominant TE mode in the rectangular waveguide (having the lowest wave attenuation).

$TM_{11}$ – Dominant TM mode in the circular waveguide (having the lowest wave attenuation).

a, b – Cross-dimensions of the rectangular waveguide in x, y coordinates.

D – Diameter of the circular waveguide.

TEM – Transversal electro-magnetic modes (in coaxial cables and coaxial waveguides).

*Microwave Plasma Sources and Methods in Processing Technology*, First Edition.
Ladislav Bárdoš and Hana Baránková.

$\lambda_g$ – The length of the propagating (guided) wave in the waveguide.

$f_{coff}$ – The critical or cut-off frequency or the lowest frequency in the waveguide.

$\lambda_{coff}$ – The critical (also $\lambda_c$) or cut-off wavelength – the longest possible in the waveguide. The dominant mode in a rectangular waveguide is $\lambda_{coff} = 2a$, in circular one is $\lambda_{coff} = 1.7$ D.

$v_{group}$ – The group velocity of the wave or the velocity of the power transfer.

$v_{phase}$ – The phase velocity or the moving wave phase in the waveguide: $v_{group} \cdot v_{phase} = c^2$.

$c_d$ – The wave propagation velocity in a dielectric; $c_d = (\mu\ \varepsilon)^{-1/2}$, where $\mu = \mu_0\ \mu_r$ and $\varepsilon = \varepsilon_0\ \varepsilon_r$; $\mu_r$ and $\varepsilon_r$ are relative permeability and permittivity of the medium.

Z – The impedance of the medium. $Z = Z_0\ (\mu/\varepsilon)^{1/2}$, where $Z_0$ is constant impedance of the free space (vacuum, also air). $Z_0 = (\mu_0/\varepsilon_0)^{1/2} = 120\ \pi = 376.73$ Ohms.

$Z_{TE}$ – The impedance of the TE mode in the waveguide: $Z_{TE} = Z_0/[1-(f_{coff}/f)^2]^{1/2}$.

$Z_{TM}$ – The impedance of the TM mode in the waveguide: $Z_{TM} = Z_0\ [1-(f_{coff}/f)^2]^{1/2}$.

SWR – Standing wave ratio, which is the ratio between highest and lowest part of the standing wave amplitude in the waveguide.

$W_r$ – Reflected power. $W_r\ [\%] = 100\ (SWR-1)^2/(SWR+1)^2$.

$\Gamma$ – The reflection coefficient: $|\Gamma| = (SWR-1)/(SWR+1)$ or $SWR = (1+|\Gamma|)/(1-|\Gamma|)$.

$P_r$ – The return loss of the power measured in decibels (dB): $P_r\ [dB] = -10 \log |\Gamma|$.

$P_{inc}$, $P_{ref}$ – The incident and reflected power; $SWR = [1+(P_{ref}/P_{inc})^{1/2}]/[1-(P_{ref}/P_{inc})^{1/2}]$.

$Z_{coax}$ – Impedance of the coaxial cable. $Z_{coax}\ [Ohm] = 60\ \varepsilon_r^{-1/2} \ln (D/d)$, where D and d are diameters of the outer and inner conductor, respectively. For $Z_{coax} = 50$ Ohms, the required ratio of diameters D/d must be: $D/d \approx \exp (0.83\ \varepsilon_r^{1/2})$.

$\Lambda$ – The length of the resonator. $\Lambda = k\ \lambda/2$ for k = 1, 2, 3, ....

tan δ – The loss tangent of material = measure of the losses by the wave absorption.

f res – The resonant frequency of cylindrical resonator: $f_{res} = [1.841^2 + (\pi r/\Lambda)^2]^{1/2}/2\pi r$, where r is the radius (r = D/2).

## Chapter 2

Temperature of 1 eV: T = eV/K = $1.6 \times 10^{-19}/1.38 \times 10^{-23}$ [As V /VAs/K] ≈ 11600 K ≈ 11300°C.

e – Electron charge (also the unit charge) e = $1.6 \times 10^{-19}$ A s. 1 A s = 1 Coulomb.

$n_e$, $n_i$ – Density of electrons, density of positive ions in the plasma ($n_e \approx n_i = n$).

Te – Electron temperature. Units are [eV] or [K]. Note 0 K = – 273.15°C. 1 eV ≈ 11600 K.

$\lambda_D$ – Debye length or plasma screening distance. $\lambda_D = e^{-1}(\varepsilon_0\, kT_e/n)^{1/2}$. Units are [m].

$\varepsilon_0$ – Permittivity of the vacuum or free-space ($\varepsilon_0 = 8.85 \times 10^{-12}$ [A s $V^{-1}m^{-1}$])

$\alpha$ – The 1st Townsend coefficient for ionization in a volume.

$\gamma$ – The 3rd Townsend coefficient = secondary electron emission coefficient = work function of the cathode metal.

$j_e$, $j_i$ – Electron current, ion current, respectively.

$V_B$ – DC breakdown voltage = voltage necessary to start the current between electrodes. The voltage depends on the gas pressure p, distance d between electrodes, and coefficients $\alpha$ and $\gamma$.

Paschen curve – Dependence $V_B$ on the pressure p and distance d.

$F_{Coulomb}$ – Coulomb force between charges $q_1$ and $q_2$. $F_{Coulomb} = k_C\,(q_1\, q_2)\,/r^2$ [Ws/m].

$k_C$ – Coulomb constant ($k_C = 9 \times 10^9$ V m/A s), r is the distance between charges.

$\omega_p$ – The plasma frequency. For electrons, $\omega_{pe} = 18\,\pi\, n^{1/2.}$

**j** – Current density vector. j = $\sigma_p$ E, which is the Ohm's law.

$\sigma_p$ – The conductivity of the plasma, which depends on the plasma frequency (plasma density) and collisions.

$\varepsilon_p$ – The permittivity of the plasma with respect to an incident electromagnetic wave with frequency $\omega$. $\varepsilon_p = \varepsilon_0\,\{1-(\omega_{pe}/\omega)^2\}$. Permittivity must be positive number = > $\omega > \omega_{pe}$.

$\nu_{coll}$ – The collision frequency between particles, mostly electrons with neutral particles in the plasma (depends on the gas pressure).

$n_{crit}$ – The critical or the cut-off density, above which the plasma starts to reflect the generating waves (power).

$\omega_{crit}$ – The critical frequency of the wave generating the critical density. The critical density for the microwave power ($f_{MW}$ = 2.4 GHz) should be $n_{crit} \approx 7 \times 10^{10}$ $cm^{-3}$.

Landau damping – The wave energy is transferred to electrons moving with a suitable phase velocity.

$r_L$ – The Larmor radius of paths of charged particles bended in the magnetic field.

ECR – Electron cyclotron resonance, when the induction $B = \omega/(e\ m_e)$.

$\omega_{ce}$ – Electron cyclotron frequency. $\omega_{ce} = eB/m_e$ or $f_{ce} = e\ B/(2\pi\ m_e)$ because $\omega_{ce} = 2\pi\ f_{ce}$.

$B_{ce}$ – The magnetic induction for the ECR. $B_{ce} = \omega\ m_e/e$.

$V_{DC}$ – The DC voltage for DC current, where $j_{DC} = \sigma_p\ E = n_e\ q_e\ v_e + n_i\ q_i\ v_i = n\ e\ (v_e + v_i)$.

$v_e$, $v_i$ – Velocity of electrons, ions. For $v_e \gg v_i$, the $j_{DC} \approx n\ e\ v_e$.

AC – Alternating voltage. It always forms an electromagnetic field.

RF – The radio frequency with a typical frequency order of $f = \omega/2\pi \approx 1–100$ MHz.

$\omega_{pi}$ – The ion plasma frequency (lower than for electrons). $\omega_{pi} = \omega_{pe}\ (m_e/m_i)^{1/2}$.

$V_{dc}$ – The DC self-bias. Found in the RF discharges due to different mobilities of electrons and ions. Self-bias depends on the frequency, electron temperature, and sizes of the RF electrodes in the plasma.

$V_s$ – Voltage across the space-charge sheath at the RF electrode. $V_s\ (t) = V_{dc} + V_{rf} \cos \omega t$.

$V_{rf}$ – Amplitude of the RF voltage applied by the RF generator.

$V_f$ – The floating potential on the unbiased electrode against to zero (ground).

$V_p$ – The plasma potential measured in the plasma against the ground.

$A_{rf}$ – The area of the RF electrode in contact with the plasma.

$A_0$ – The area of the grounded electrode (or grounded parts in the reactor in contact with the plasma. The relation is $V_s/V_p = (A_0/A_{rf})^2$.

$d_s$ – The thickness of the space-charge sheath at the electrode. $d_s \approx (1/\text{pressure})^{1/2}$.

Thermal plasma – The plasma where all particles have the same energy (in thermal equilibrium) $\leq >$ hot plasma.

Cold plasma – The plasma where different particles have different energies. Also non-thermal or non-equilibrium plasma. The small electrons in such plasmas have highest energies.

$n_{gas}$ – The density of neutral particles in the plasma. It depends on the pressure and temperature T and can be estimated from $n_{gas}\ [m^{-3}] = 7.2\ 10^{22}\ p\ [Pa]/(T[°C] + 273) = 9.6\ 10^{24}\ p\ [Torr]/(T[°C] + 273) = 7.2\ 10^{24}\ p\ [mBar]/(T[°C] + 273)$.

$\delta_i$ – The degree of the ionization in a given plasma. $\delta_i = n/n_{gas}$.

$l_p$ – The mean-free path (MFP). The path of the particle between collisions.

$Q_{p\text{-}g}$ – The collision cross-section of a gas particle characteristic for the given gas.

$E_a$ – The activation energy ($kT_a$) of a given chemical reaction in equilibrium chemistry. (k is the Boltzmann constant.)

Tg, Tt, Tr, – Energies (temperatures) of the respective individual
Tv, Tel – particles in the plasma: gas, translation (heat), rotation (atoms in molecules), vibration (atoms in molecules), and electrons orbits in atoms (molecules).

$W_e$ – Electron energy in the plasma. $W_e \approx kT_e$.

EEDF – Electron energy distribution function, i.e. the number of electrons in each energy level (from 0 to infinity). The two most frequent ones are Maxwellian and Druywestain
EEDFs.

$f_{MW}$ – The microwave frequency. Typically, $f_{MW} = 2.4 \pm 0.05$ GHz. The critical density for this wave frequency is $n_{crit} \approx 7 \times 10^{10}$ $cm^{-3}$.

## Chapter 3

$p_{TOT}$ – Total gas pressure, consisting of partial pressures of the participating gases in the mixture $p_A + p_B + p_C + .$

$n_{gas}$, n, $n_e$, $n_i^-$, – The particle densities of the neutral gas, the plasma density,
$n_i^+$ density of electrons, negative ions, and positive ions, respectively. They are related as $n = n_e + (n_i^-) = n_i^+$.

$\delta_i$ – The degree of the ionization in given plasma. $\delta_i = n/n_{gas}$.

$N_e$ – The electron density. Used sometimes instead of $n_e$.

CVD – Chemical Vapor Deposition – deposition of coatings from gaseous phases.

PVD – Physical Vapor Deposition – deposition from evaporated or sputtered solid targets.

PE CVD – Plasma-Enhanced CVD. Also Plasma-Assisted CVD (PA CVD), Plasma-Activated CVD (PA CVD), or Plasma-Induced CVD (PI CVD).

RIE – Reactive ion etching. Ion-enhanced chemical etching using plasma.

$k_B$ – The Boltzmann constant. $k_B = k$.

$m_{gas}$ – Mass of gas particles in the plasma.

$n_{gas}$ – The density of neutral particles in the plasma.

$\mathbf{F_D}$ – The force vector forcing particles from higher to lower density. $\mathbf{F_D} \approx$ grad n (x, y, z).

$\Phi$ – Flux of the diffused neutral particles. $\Phi = D\, n(x)/dx$.

D – Diffusion coefficient.

$N_{gas}$ – Total number of particles in a gas volume ($V_{gas}$): $N_{gas} = n_{gas}.\ V_{gas}$. Here, $p_{gas}.V_{gas} = N_{gas}\, k_B T_{gas}$ (the general law for an ideal gas).

$V_{gas}$ – The volume occupied by gas particles in a reactor vessel.

$p_{gas}$ – The gas pressure in the reactor vessel.

$\mathbf{F_{e,i}}$ – The Lorentz force vector acting on charged particles in electric and magnetic fields ($q_{e,i}$ are charge of electrons and/or ions). $F_{e,i} = q_{e,i}\, E + q_{e,i}\, (v_{e,i} \times B)$ [N, C, $m.s^{-1}$, T].

$q_{e,i}$ – Charges. For single ionized ions, $|q_i| = |q_e| = e = 1.6 \times 10^{-19}$ A.s.

$r_L$ – Larmor radius of charges in a magnetic field with induction B (for charges moving perpendicularly to the vector B). $r_L = m_{e,i}\, v_{norm\,(e,i)} / (|q_{e,i}|\, B)$.

$\mathbf{F_B}$ – Force vector acting on charges in non-uniform field B. $\mathbf{F_B} \approx$ grad B (x,y,z).

$\mu_p$ – The mobility of particles having mass m moving in electric field E with the drift velocity $v_d$ – It is expressed as $\mu_p = v_d/E = q/(m\, \nu_m)$.

$\nu_m$ – The collision frequency with the transfer of the particle momentum.

$m_i$ – Mass of ions (about 1000 times higher than mass of electrons). The mass of a proton is $1.67 \times 10^{-27\,kg}$ The mass of an ion depends on the content of its protons.

$\nu_{coll\,(e,i\,-\,gas)}$ – The collision frequency of electrons and ions with neutral particles. The prevailing colissions are between electrons and neutral gas particles, because electrons have highest mobility.

$T_c$ – Period of cycles in pulsing.

Duty-cycle – The time ratio of the pulse to the pause between pulses.

CW – Continual Wave = continual power without pulsing.

## Chapter 4

$B_{CE}$ – The magnetic induction necessary for the ECR (resonance) of the incident microwaves. The resonance condition for wave frequency $f = \omega/2\pi$ is $B = B_{ce}$ (or $\omega_{ce}/\omega = 1$).

MW – A sign used for the microwave generator or power. Also μW or μw.

Thickness – The thickness units for coatings or films or depths in an oxidized surface are [nm] or [μm]. For thin thickness, use [Å]. 1 μm = 1000 nm = 10 000 Å.

$U_{fl}$ – The floating potential of an unbiased and ungrounded electrode in the plasma. It is the bias (with respect to the ground) when the fluxes of negative electrons and positive ions are compensated and the resulting current to the electrode is zero.

$A_p$ – An effective area of the probe for electrical measurements in the plasma.

$I_s$ – Saturated ion current. It is the current limited by an available plasma density.

$V_{probe}$ – Electric potential (with respect to the ground) applied on the probe for measurements in the plasma.

N – The plasma density (also n). The density number in unit volume ($m^{-3}$ or $cm^{-3}$) is the same for electrons and ions.

$E_e$ – The electron energy in eV.

$T_{sub}$ – Substrate temperature. The temperature of the sample in plasma process.

L – Length of a resonator.

Langmuir probe – (commercially available devices) – the thin wire (single or double) biased electrically or unbiased (floating) and immersed in the plasma for measurements of local plasma density, electron energy, floating potential, plasma potential and for the electron energy distribution function (only single probe). The probe measurements are invasive and require often complicated interpretations. The probes are used at pressures below about 10 Torr. For details, see, e.g. F.F. Chen and J.P. Chang: "Langmuir Probes" – Chapter XII in "Lecture Notes on Principles of Plasma Processing"; Springer Science + Business Media, LLC, (2003) ISBN 978-1-4615-0181-7 (eBook). DOI 10.1007/978-1-4615-0181-7. https://link.springer.com/book/10.1007/978-1-4615-0181-7.

## Chapter 5

$V_B$ – The breakdown voltage of gas. The voltage needed to start the plasma.

DBD – Dielectric Barrier Discharge = an atmospheric discharge between electrodes shielded by a dielectric barrier to prevent arcing. The dielectric shield can be on one or both electrodes.

H-HEAD – Hybrid Hollow Electrode Activated Discharge.

LPG – Liquified Petroleum Gas = mixes of propane ($C_3H_8$) and butane ($C_4H_{10}$).

## 7.2 Constants and Numbers

**Electron charge** $e = 1.6 \times 10^{-19}$ A s.

**Electron mass** $m_e = 9.11 \times 10^{-31}$ kg.

**Ion mass** – dependent of number of protons (proton mass is $1.67 \times 10^{-27}$ kg).

**Diameter of an atom** – in the range of $1 \times 10^{-10}$ m (0.1 nm = 1 Å).

**Diameter of the nucleus** – in the range of $1.75 \times 10^{-15}$ m (1.75 fm). (The size ratio atom/nucleus $\approx 10^5$. It is same as the soccer stadium/mosquito $\approx$ 100 m/1 mm $= 10^5$).

**Diameter of an electron** –$5.64 \times 10^{-15}$ m (Pauling 1964). More recent data show $< 10^{-18}$ m.

**Boltzmann constant** $k = 1.38 \times 10^{-23}$ W s/K.

**Planck constant** $h = 6.63 \times 10^{-34}$ kg m²/s.

**1 atm** = 1.01 Bar = 760 Torr = $1.01 \times 10^5$ Pa, 1 mBar = 100 Pa.

**1 Torr** = 133.3 Pa = 1.33 mBar —> $3 \times 10^{16}$molecules/cm³ (at 0°C).

**1 Pa** = $7.5 \times 10^{-3}$ Torr, 1 mBar = 0.75 Torr.

**Avogadro number** $N_A = 6.022 \times 10^{23}$ mol$^{-1}$.

**sccm** = flow of the standard cubic centimeter per minute = flow of 1 cm³ at standard pressure of 1 atm.

**slm** = flow of the standard liter per minute = 1000 sccm.

**1 sccm** = $6.022 \times 10^{23}/22.414$ molec/min = $2.69 \times 10^{19}$ molecules/min.

**1 Torr litre/sec** = 79.03 sccm = $2.13 \times 10^{21}$ molecules/min.

**Molar volume:** $V_{mol} = 22.41 \times 10^{-3}$ m³/mol.

**Mean free path** (20°C) at 1 mTorr ≈ 50 mm.

**Pressure intervals** < 1 mTorr (<0.13 Pa) => less than $10^{13}$ particles in cm³, 10 mTorr-10 Torr (1.3 Pa – $1.33 \times 10^3$ Pa) => $10^{14}$ – $10^{17}$ particles/cm³, > 10 Torr (> $1.33 \times 10^3$ Pa) => more than $10^{17}$ particles/cm³.

**Vacuum permittivity:** $\varepsilon_o = 8.85 \times 10^{-12}$ Farad/m [As/Vm].

**Vacuum permeability:** $\mu_o = 1.26 \times 10^{-6}$ Henry/m [Vs/Am].

**An impedance for wave propagation in the free space:** $Z_{free\ space} = (\mu_o/\varepsilon_o)^{1/2} = 377\ \Omega$.

# Index



*Microwave Plasma Sources and Methods in Processing Technology*, First Edition.
Ladislav Bárdoš and Hana Baránková.

## e



## f



## g



## h



## i